Honda MBX/MTX 125 & MTX200 Owners Workshop Manual

by Jeremy Churchill

Models covered

MBX125 F. 125cc. January 1984 to September 1985
MTX125 RW. 125cc. February 1983 on
MTX200 RW. 194cc. March 1983 to September 1986

(1132 - 12R7)

J H Haynes & Co. Ltd.
Haynes North America, Inc

www.haynes.com

Acknowledgements

Our thanks are due to Paul Branson Motorcycles of Yeovil who supplied the MBX125 model, Franks Autoneeds and Cycles of Langport who supplied the MTX125 RW-D model, and Atkins Motors of Taunton who supplied the MTX125 RW-L model. We would also like to thank the Avon Rubber Company, who kindly supplied information and technical assistance on tyre fitting; NGK Spark Plugs (UK) Ltd for information on spark plug maintenance and electrode conditions, and Renold Ltd for advice on chain care and renewal.

A book in the **Haynes Owners Workshop Manual Series**

ISBN 978 1 85010 901 3

British Library Cataloguing in Publication Data

A catalogue record for this book is available from the British Library

Contents

Left-hand view of the Honda MBX125 F-E

Left-hand view of the Honda MTX125 RW-D

About this manual

The purpose of this manual is to present the owner with a concise and graphic guide which will enable him to tackle any operation from basic routine maintenance to a major overhaul. It has been assumed that any work would be undertaken without the luxury of a well-equipped workshop and a range of manufacturer's service tools.

To this end, the machine featured in the manual was stripped and rebuilt in our own workshop, by a team comprising a mechanic, a photographer and the author. The resulting photographic sequence depicts events as they took place, the hands shown being those of the author and the mechanic.

The use of specialised, and expensive, service tools was avoided unless their use was considered to be essential due to risk of breakage or injury. There is usually some way of improvising a method of removing a stubborn component, providing that a suitable degree of care is exercised.

The author learnt his motorcycle mechanics over a number of years, faced with the same difficulties and using similar facilities to those encountered by most owners. It is hoped that this practical experience can be passed on through the pages of this manual.

Where possible, a well-used example of the machine is chosen for the workshop project, as this highlights any areas which might be particularly prone to giving rise to problems. In this way, any such difficulties are encountered and resolved before the text is written, and the techniques used to deal with them can be incorporated in the relevant section. Armed with a working knowledge of the machine, the author undertakes a considerable amount of research in order that the maximum amount of data can be included in the manual.

A comprehensive section, preceding the main part of the manual, describes procedures for carrying out the routine maintenance of the machine at intervals of time and mileage. This section is included particularly for those owners who wish to ensure the efficient day-to-day running of their motorcycle, but who choose not to undertake overhaul or renovation work.

Each Chapter is divided into numbered sections. Within these sections are numbered paragraphs. Cross reference throughout the manual is quite straightforward and logical. When reference is made 'See Section 6.10' it means Section 6, paragraph 10 in the same Chapter. If another Chapter were intended, the reference would read, for example, 'See Chapter 2, Section 6.10'. All the photographs are captioned with a section/paragraph number to which they refer and are relevant to the Chapter text adjacent.

Figures (usually line illustrations) appear in a logical but numerical order, within a given Chapter. Fig. 1.1 therefore refers to the first figure in Chapter 1.

Left-hand and right-hand descriptions of the machines and their components refer to the left and right of a given machine when the rider is seated normally.

Motorcycle manufacturers continually make changes to specifications and recommendations, and these, when notified, are incorporated into our manuals at the earliest opportunity.

Introduction to the Honda MBX/MTX125 and MTX200 models

Before 1982 the 125cc capacity class of motorcycles was not particularly popular in the UK, since the machines were too slow to attract in large numbers the younger riders looking for performance above all else, and yet were too large to be attractive to the commuter, or non-enthusiast rider. With a few notable exceptions, principally in the trail bike class, their relatively dull specifications reflected the manufacturers natural lack of interest.

This situation was altered radically by the legislation that came into force during 1982 and early 1983; all learner motorcyclists were to be restricted to machines of a maximum engine size of 125cc, the power output being restricted to 9kW (12.2 bhp). Almost immediately the four Japanese manufacturers responded with a confusing mass of new models designed to attract the new captive market. Features formerly considered worthwhile only on machines of much larger engine capacity were included in the specification of the new models, many of which incorporated the latest ideas in suspension and styling.

Honda's new models were announced in 1983, the two MTX models appearing first, to be followed by the MBX model in early 1984. In that they feature two-stroke engines the new machines were something of a departure for a factory normally known for four-strokes. However Honda have been using two-strokes more frequently in recent years, especially in competition, and it is only natural that they should appear in the range offered for sale to the general public.

The MBX125 is designed and equipped as a road-going sports model with its Comstar wheels, hydraulic disc brake, miniature fairing and general styling; all features that are now becoming standard equipment in its class. Since the restrictions imposed mean that there can be little to choose between the different models as far as performance is concerned, the manufacturers have to ensure that their machines are as well-equipped and as visually attractive as possible to ensure good sales; in this Honda would appear to have succeeded.

The MTX models are designed to replace Honda's well-known XL/XR series of models in the trail bike class and are thus equipped with all the latest features and styling trends that the customer has come to expect; with their long-travel suspension and very striking appearance these models echo the success of the CR series of motocross machines.

Although the two 125cc models are detuned to comply with UK legislation, the MTX200 engine is claimed to have a power output considerably above that of the XR200 four-stroke model that it replaces and should prove popular with the rider who wants a competitive machine without the extra size, cost and weight of a larger capacity model. At the time of writing, although unrestricted 125cc models exist for sale in other countries, they are not sold in the UK.

While the MBX125 F-E model remained unchanged until it was discontinued in September 1985, the MTX125/200 models appeared in new versions. The original MTX125/200 RW-D models were fitted with drum front brakes and had the pillion footrests mounted on the swinging arm. These were replaced by the MTX125/200 RW-F models during 1985, although the MTX200 RW-F was never imported into the UK. The RW-F models can be easily distinguished by their hydraulically-operated disc front brakes. In addition to this they had revised front and rear suspension components and the pillion footrests were mounted on separate frame-mounted brackets, as well as changes to paintwork, graphics and other minor components. With the 200 model being phased out, the 125 was replaced by the MTX125 RW-H in late 1987 and then the MTX125 RW-L in late 1989; these models differ in detail only, particularly changes to the paintwork and graphics.

To help owners identify their machines exactly, the approximate dates of import are given below, with the initial engine and frame number with which each model's production run commenced.

Model	Date	Engine number	Frame number
MBX125 F-E	Jan '84 to Sept '85	N/Av	N/Av
MTX125 RW-D	Feb '83 to Apr '85	JD05E-5000033 on	JD05-5000034 on
MTX125 RW-F	Apr '85 to Nov '87	JD05E-5016283 on	JD05-5101019 on
MTX125 RW-H	Oct '87 to Dec '89	JD05E-5100001 on	JD05-5200001 on
MTX125 RW-L	Dec '89 on	JD05E-5024787 on	JD05-5300001 on
MTX200 RW-D	Mar '83 to Sept '86	JD07E-5000031 on	MD07-5000030 on
MTX200 RW-F	N/Av (not imported into UK)		

Model dimensions and weights

	MBX125 F-E	MTX125/200 RW-D	MTX125 RW-F, H, L, MTX200 RW-F
Overall length	1970 mm (77.6 in)	2090 mm (82.3 in)	2090 mm (82.3 in)
Overall width	700 mm (27.6 in)	830 mm (32.7 in)	830 mm (32.7 in)
Overall height	1110 mm (43.7 in)	1175 mm (46.3 in)	1175 mm (46.3 in)
Wheelbase	1310 mm (51.6 in)	1345 mm (53.0 in)	1345 mm (53.0 in)
Seat height	760 mm (29.9 in)	845 mm (33.3 in)	870 mm (34.3 in)
Ground clearance	160 mm (6.3 in)	285 mm (11.2 in)	295 mm (11.6 in)
Dry weight	99 kg (218 lb)	99 kg (218 lb) – 125 101 kg (223 lb) – 200	103 kg (227 lb) – 125 104 kg (229 lb) – 200
Kerb weight	112 kg (247 lb)	112 kg (247 lb) – 125 114 kg (251 lb) – 200	116 kg (256 lb) – 125 117 kg (258 lb) – 200

Ordering spare parts

When ordering spare parts it is advisable to deal direct with an official Honda agent, who will be able to supply many of the items required ex-stock. It is advisable to get acquainted with the local Honda agent, and to rely on his advice when purchasing spares. He is in a better position to specify exactly the parts required and to identify the relevant spare part numbers so that there is less chance of the wrong parts being supplied by the manufacturer due to a vague or incomplete description.

When ordering spares, always quote the frame and engine numbers in full, together with any prefixes or suffixes in the form of letters. The frame number is found stamped on the right-hand side of the steering head, in line with the forks. The engine number is stamped on the left-hand side of the crankcase, immediately in front of the gearchange lever shaft.

Use only parts of genuine Honda manufacture. A few pattern parts are available, sometimes at cheaper prices, but there is no guarantee that they will give such good service as the originals they replace. Retain any worn or broken parts until the replacements have been obtained; they are sometimes needed as a pattern to help identify the correct replacement when design changes have been made during a production run.

Some of the more expendable parts such as spark plugs, bulbs, tyres, oils and greases etc., can be obtained from accessory shops and motor factors, who have convenient opening hours, and can often be found not far from home. It is also possible to obtain parts on a Mail Order basis from a number of specialists who advertise regularly in the motorcycle magazines.

Location of frame number

Location of engine number

Safety first!

Professional motor mechanics are trained in safe working procedures. However enthusiastic you may be about getting on with the job in hand, do take the time to ensure that your safety is not put at risk. A moment's lack of attention can result in an accident, as can failure to observe certain elementary precautions.

There will always be new ways of having accidents, and the following points do not pretend to be a comprehensive list of all dangers; they are intended rather to make you aware of the risks and to encourage a safety-conscious approach to all work you carry out on your vehicle.

Essential DOs and DON'Ts

DON'T start the engine without first ascertaining that the transmission is in neutral.

DON'T suddenly remove the filler cap from a hot cooling system – cover it with a cloth and release the pressure gradually first, or you may get scalded by escaping coolant.

DON'T attempt to drain oil until you are sure it has cooled sufficiently to avoid scalding you.

DON'T grasp any part of the engine, exhaust or silencer without first ascertaining that it is sufficiently cool to avoid burning you.

DON'T allow brake fluid or antifreeze to contact the machine's paintwork or plastic components.

DON'T syphon toxic liquids such as fuel, brake fluid or antifreeze by mouth, or allow them to remain on your skin.

DON'T inhale dust – it may be injurious to health (see *Asbestos* heading).

DON'T allow any spilt oil or grease to remain on the floor – wipe it up straight away, before someone slips on it.

DON'T use ill-fitting spanners or other tools which may slip and cause injury.

DON'T attempt to lift a heavy component which may be beyond your capability – get assistance.

DON'T rush to finish a job, or take unverified short cuts.

DON'T allow children or animals in or around an unattended vehicle.

DON'T inflate a tyre to a pressure above the recommended maximum. Apart from overstressing the carcase and wheel rim, in extreme cases the tyre may blow off forcibly.

DO ensure that the machine is supported securely at all times. This is especially important when the machine is blocked up to aid wheel or fork removal.

DO take care when attempting to slacken a stubborn nut or bolt. It is generally better to pull on a spanner, rather than push, so that if slippage occurs you fall away from the machine rather than on to it.

DO wear eye protection when using power tools such as drill, sander, bench grinder etc.

DO use a barrier cream on your hands prior to undertaking dirty jobs – it will protect your skin from infection as well as making the dirt easier to remove afterwards; but make sure your hands aren't left slippery. Note that long-term contact with used engine oil can be a health hazard.

DO keep loose clothing (cuffs, tie etc) and long hair well out of the way of moving mechanical parts.

DO remove rings, wristwatch etc, before working on the vehicle – especially the electrical system.

DO keep your work area tidy – it is only too easy to fall over articles left lying around.

DO exercise caution when compressing springs for removal or installation. Ensure that the tension is applied and released in a controlled manner, using suitable tools which preclude the possibility of the spring escaping violently.

DO ensure that any lifting tackle used has a safe working load rating adequate for the job.

DO get someone to check periodically that all is well, when working alone on the vehicle.

DO carry out work in a logical sequence and check that everything is correctly assembled and tightened afterwards.

DO remember that your vehicle's safety affects that of yourself and others. If in doubt on any point, get specialist advice.

IF, in spite of following these precautions, you are unfortunate enough to injure yourself, seek medical attention as soon as possible.

Asbestos

Certain friction, insulating, sealing, and other products – such as brake linings, clutch linings, gaskets, etc – contain asbestos. *Extreme care must be taken to avoid inhalation of dust from such products since it is hazardous to health.* If in doubt, assume that they *do* contain asbestos.

Fire

Remember at all times that petrol (gasoline) is highly flammable. Never smoke, or have any kind of naked flame around, when working on the vehicle. But the risk does not end there – a spark caused by an electrical short-circuit, by two metal surfaces contacting each other, by careless use of tools, or even by static electricity built up in your body under certain conditions, can ignite petrol vapour, which in a confined space is highly explosive.

Always disconnect the battery earth (ground) terminal before working on any part of the fuel or electrical system, and never risk spilling fuel on to a hot engine or exhaust.

It is recommended that a fire extinguisher of a type suitable for fuel and electrical fires is kept handy in the garage or workplace at all times. Never try to extinguish a fuel or electrical fire with water.

Note: *Any reference to a 'torch' appearing in this manual should always be taken to mean a hand-held battery-operated electric lamp or flashlight. It does **not** mean a welding/gas torch or blowlamp.*

Fumes

Certain fumes are highly toxic and can quickly cause unconsciousness and even death if inhaled to any extent. Petrol (gasoline) vapour comes into this category, as do the vapours from certain solvents such as trichloroethylene. Any draining or pouring of such volatile fluids should be done in a well ventilated area.

When using cleaning fluids and solvents, read the instructions carefully. Never use materials from unmarked containers – they may give off poisonous vapours.

Never run the engine of a motor vehicle in an enclosed space such as a garage. Exhaust fumes contain carbon monoxide which is extremely poisonous; if you need to run the engine, always do so in the open air or at least have the rear of the vehicle outside the workplace.

The battery

Never cause a spark, or allow a naked light, near the vehicle's battery. It will normally be giving off a certain amount of hydrogen gas, which is highly explosive.

Always disconnect the battery earth (ground) terminal before working on the fuel or electrical systems.

If possible, loosen the filler plugs or cover when charging the battery from an external source. Do not charge at an excessive rate or the battery may burst.

Take care when topping up and when carrying the battery. The acid electrolyte, even when diluted, is very corrosive and should not be allowed to contact the eyes or skin.

If you ever need to prepare electrolyte yourself, always add the acid slowly to the water, and never the other way round. Protect against splashes by wearing rubber gloves and goggles.

Mains electricity and electrical equipment

When using an electric power tool, inspection light etc, always ensure that the appliance is correctly connected to its plug and that, where necessary, it is properly earthed (grounded). Do not use such appliances in damp conditions and, again, beware of creating a spark or applying excessive heat in the vicinity of fuel or fuel vapour. Also ensure that the appliances meet the relevant national safety standards.

Ignition HT voltage

A severe electric shock can result from touching certain parts of the ignition system, such as the HT leads, when the engine is running or being cranked, particularly if components are damp or the insulation is defective. Where an electronic ignition system is fitted, the HT voltage is much higher and could prove fatal.

Tools and working facilities

The first priority when undertaking maintenance or repair work of any sort on a motorcycle is to have a clean, dry, well-lit working area. Work carried out in peace and quiet in the well-ordered atmosphere of a good workshop will give more satisfaction and much better results than can usually be achieved in poor working conditions. A good workshop must have a clean flat workbench or a solidly constructed table of convenient working height. The workbench or table should be equipped with a vice which has a jaw opening of at least 4 in (100 mm). A set of jaw covers should be made from soft metal such as aluminium alloy or copper, or from wood. These covers will minimise the marking or damaging of soft or delicate components which may be clamped in the vice. Some clean, dry, storage space will be required for tools, lubricants and dismantled components. It will be necessary during a major overhaul to lay out engine/gearbox components for examination and to keep them where they will remain undisturbed for as long as is necessary. To this end it is recommended that a supply of metal or plastic containers of suitable size is collected. A supply of clean, lint-free, rags for cleaning purposes and some newspapers, other rags, or paper towels for mopping up spillages should also be kept. If working on a hard concrete floor note that both the floor and one's knees can be protected from oil spillages and wear by cutting open a large cardboard box and spreading it flat on the floor under the machine or workbench. This also helps to provide some warmth in winter and to prevent the loss of nuts, washers, and other tiny components which have a tendency to disappear when dropped on anything other than a perfectly clean, flat, surface.

Unfortunately, such working conditions are not always available to the home mechanic. When working in poor conditions it is essential to take extra time and care to ensure that the components being worked on are kept scrupulously clean and to ensure that no components or tools are lost or damaged.

A selection of good tools is a fundamental requirement for anyone contemplating the maintenance and repair of a motor vehicle. For the owner who does not possess any, their purchase will prove a considerable expense, offsetting some of the savings made by doing-it-yourself. However, provided that the tools purchased meet the relevant national safety standards and are of good quality, they will last for many years and prove an extremely worthwhile investment.

To help the average owner to decide which tools are needed to carry out the various tasks detailed in this manual, we have compiled three lists of tools under the following headings: *Maintenance and minor repair, Repair and overhaul,* and *Specialized.* The newcomer to practical mechanics should start off with the simpler jobs around the vehicle. Then, as his confidence and experience grow, he can undertake more difficult tasks, buying extra tools as and when they are needed. In this way, a *Maintenance and minor repair* tool kit can be built-up into a *Repair and overhaul* tool kit over a considerable period of time without any major cash outlays. The experienced home mechanic will have a tool kit good enough for most repair and overhaul procedures and will add tools from the specialized category when he feels the expense is justified by the amount of use these tools will be put to.

It is obviously not possible to cover the subject of tools fully here. For those who wish to learn more about tools and their use there is a book entitled *Motorcycle Workshop Practice Manual* available from the publishers of this manual. It also provides an introduction to basic workshop practice which will be of interest to a home mechanic working on any type of motor vehicle.

As a general rule, it is better to buy the more expensive, good quality tools. Given reasonable use, such tools will last for a very long time, whereas the cheaper, poor quality, item will wear out faster and need to be renewed more often, thus nullifying the original saving. There is also the risk of a poor quality tool breaking while in use, causing personal injury or expensive damage to the component being worked on.

For practically all tools, a tool factor is the best source since he will have a very comprehensive range compared with the average garage or accessory shop. Having said that, accessory shops often offer excellent quality tools at discount prices, so it pays to shop around. There are plenty of tools around at reasonable prices, but always aim to purchase items which meet the relevant national safety standards. If in doubt, seek the advice of the shop proprietor or manager before making a purchase.

The basis of any toolkit is a set of spanners. While open-ended spanners with their slim jaws, are useful for working on awkwardly-positioned nuts, ring spanners have advantages in that they grip the nut far more positively. There is less risk of the spanner slipping off the nut and damaging it, for this reason alone ring spanners are to be preferred. Ideally, the home mechanic should acquire a set of each, but if expense rules this out a set of combination spanners (open-ended at one end and with a ring of the same size at the other) will provide a good compromise. Another item which is so useful it should be considered an essential requirement for any home mechanic is a set of socket spanners. These are available in a variety of drive sizes. It is recommended that the $1/2$-inch drive type is purchased to begin with as although bulkier and more expensive than the $3/8$-inch type, the larger size is far more common and will accept a greater variety of torque wrenches, extension pieces and socket sizes. The socket set should comprise sockets of sizes between 8 and 24 mm, a reversible ratchet drive, an extension bar of about 10 inches in length, a spark plug socket with a rubber insert, and a universal joint. Other attachments can be added to the set at a later date.

Maintenance and minor repair tool kit

Set of spanners 8 – 24 mm
Set of sockets and attachments
Spark plug spanner with rubber insert – 10, 12, or 14 mm as appropriate
Adjustable spanner
C-spanner/pin spanner
Torque wrench (same size drive as sockets)
Set of screwdrivers (flat blade)
Set of screwdrivers (cross-head)
Set of Allen keys 4 – 10 mm
Impact screwdriver and bits
Ball pein hammer – 2 lb
Hacksaw (junior)
Self-locking pliers – Mole grips or vice grips
Pliers – combination
Pliers – needle nose
Wire brush (small)
Soft-bristled brush
Tyre pump
Tyre pressure gauge
Tyre tread depth gauge
Oil can
Fine emery cloth
Funnel (medium size)
Drip tray
Grease gun
Set of feeler gauges
Brake bleeding kit
Strobe timing light
Continuity tester (dry battery and bulb)
Soldering iron and solder
Wire stripper or craft knife
PVC insulating tape
Assortment of split pins, nuts, bolts, and washers

Repair and overhaul toolkit

The tools in this list are virtually essential for anyone undertaking major repairs to a motorcycle and are additional to the tools listed above. Concerning Torx driver bits, Torx screws are encountered on some of the more modern machines where their use is restricted to fastening certain components inside the engine/gearbox unit. It is therefore recommended that if Torx bits cannot be borrowed from a local dealer, they are purchased individually as the need arises. They are not in regular use in the motor trade and will therefore only be available in specialist tool shops.

Plastic or rubber soft-faced mallet
Torx driver bits
Pliers – electrician's side cutters
Circlip pliers – internal (straight or right-angled tips are available)
Circlip pliers – external
Cold chisel
Centre punch
Pin punch

Scriber
Scraper (made from soft metal such as aluminium or copper)
Soft metal drift
Steel rule/straight edge
Assortment of files
Electric drill and bits
Wire brush (large)
Soft wire brush (similar to those used for cleaning suede shoes)
Sheet of plate glass
Hacksaw (large)
Stud extractor set (E-Z out)

Specialized tools

This is not a list of the tools made by the machine's manufacturer to carry out a specific task on a limited range of models. Occasional references are made to such tools in the text of this manual and, in general, an alternative method of carrying out the task without the manufacturer's tool is given where possible. The tools mentioned in this list are those which are not used regularly and are expensive to buy in view of their infrequent use. Where this is the case it may be possible to hire or borrow the tools against a deposit from a local dealer or tool hire shop. An alternative is for a group of friends or a motorcycle club to join in the purchase.

Piston ring compressor
Universal bearing puller
Cylinder bore honing attachment (for electric drill)
Micrometer set
Vernier calipers
Dial gauge set
Cylinder compression gauge
Multimeter
Dwell meter/tachometer

Care and maintenance of tools

Whatever the quality of the tools purchased, they will last much longer if cared for. This means in practice ensuring that a tool is used for its intended purpose; for example screwdrivers should not be used as a substitute for a centre punch, or as chisels. Always remove dirt or grease and any metal particles but remember that a light film of oil will prevent rusting if the tools are infrequently used. The common tools can be kept together in a large box or tray but the more delicate, and more expensive, items should be stored separately where they cannot be damaged. When a tool is damaged or worn out, be sure to renew it immediately. It is false economy to continue to use a worn spanner or screwdriver which may slip and cause expensive damage to the component being worked on.

Fastening systems

Fasteners, basically, are nuts, bolts and screws used to hold two or more parts together. There are a few things to keep in mind when working with fasteners. Almost all of them use a locking device of some type; either a lock washer, lock nut, locking tab or thread adhesive. All threaded fasteners should be clean, straight, have undamaged threads and undamaged corners on the hexagon head where the spanner fits. Develop the habit of replacing all damaged nuts and bolts with new ones.

Rusted nuts and bolts should be treated with a rust penetrating fluid to ease removal and prevent breakage. After applying the rust penetrant, let it 'work' for a few minutes before trying to loosen the nut or bolt. Badly rusted fasteners may have to be chiseled off or removed with a special nut breaker, available at tool shops.

Flat washers and lock washers, when removed from an assembly should always be replaced exactly as removed. Replace any damaged washers with new ones. Always use a flat washer between a lock washer and any soft metal surface (such as aluminium), thin sheet metal or plastic. Special lock nuts can only be used once or twice before they lose their locking ability and must be renewed.

If a bolt or stud breaks off in an assembly, it can be drilled out and removed with a special tool called an E-Z out. Most dealer service departments and motorcycle repair shops can perform this task, as well as others (such as the repair of threaded holes that have been stripped out).

Spanner size comparison

Spanner size comparison

Jaw gap (in)	Spanner size	Jaw gap (in)	Spanner size
0.250	1/4 in AF	0.945	24 mm
0.276	7 mm	1.000	1 in AF
0.313	5/16 in AF	1.010	9/16 in Whitworth; 5/8 in BSF
0.315	8 mm	1.024	26 mm
0.344	11/32 in AF; 1/8 in Whitworth	1.063	11/16 in AF; 27 mm
0.354	9 mm	1.100	5/16 in Whitworth; 11/16 in BSF
0.375	3/8 in AF	1.125	11/8 in AF
0.394	10 mm	1.181	30 mm
0.433	11 mm	1.200	11/16 in Whitworth; 3/4 in BSF
0.438	7/16 in AF	1.250	11/4 in AF
0.445	3/16 in Whitworth; 1/4 in BSF	1.260	32 mm
0.472	12 mm	1.300	3/4 in Whitworth; 7/8 in BSF
0.500	1/2 in AF	1.313	15/16 in AF
0.512	13 mm	1.390	13/16 in Whitworth; 15/15 in BSF
0.525	1/4 in Whitworth; 5/16 in BSF	1.417	36 mm
0.551	14 mm	1.438	17/16 in AF
0.563	9/16 in AF	1.480	7/8 in Whitworth; 1 in BSF
0.591	15 mm	1.500	11/2 in AF
0.600	5/16 in Whitworth; 3/8 in BSF	1.575	40 mm; 15/16 in Whitworth
0.625	5/8 in AF	1.614	41 mm
0.630	16 mm	1.625	15/8 in AF
0.669	17 mm	1.670	1 in Whitworth; 11/8 in BSF
0.686	11/16 in AF	1.688	111/16 in AF
0.709	18 mm	1.811	46 mm
0.710	3/8 in Whitworth; 7/16 in BSF	1.813	113/16 in AF
0.748	19 mm	1.860	11/8 in Whitworth; 11/4 in BSF
0.750	3/4 in AF	1.875	17/8 in AF
0.813	13/16 in AF	1.969	50 mm
0.820	7/16 in Whitworth; 1/2 in BSF	2.000	2 in AF
0.866	22 mm	2.050	11/4 in Whitworth; 13/8 in BSF
0.875	7/8 in AF	2.165	55 mm
0.920	1/2 in Whitworth; 9/16 in BSF	2.362	60 mm
0.938	15/16 in AF		

Standard torque settings

Specific torque settings will be found at the end of the specifications section of each chapter. Where no figure is given, bolts should be secured according to the table below.

Fastener type (thread diameter)	kgf m	lbf ft
5mm bolt or nut	0.45 – 0.6	3.5 – 4.5
6 mm bolt or nut	0.8 – 1.2	6 – 9
8 mm bolt or nut	1.8 – 2.5	13 – 18
10 mm bolt or nut	3.0 – 4.0	22 – 29
12 mm bolt or nut	5.0 – 6.0	36 – 43
5 mm screw	0.35 – 0.5	2.5 – 3.6
6 mm screw	0.7 – 1.1	5 – 8
6 mm flange bolt	1.0 – 1.4	7 – 10
8 mm flange bolt	2.4 – 3.0	17 – 22
10 mm flange bolt	3.0 – 4.0	22 – 29

Choosing and fitting accessories

The range of accessories available to the modern motorcyclist is almost as varied and bewildering as the range of motorcycles. This Section is intended to help the owner in choosing the correct equipment for his needs and to avoid some of the mistakes made by many riders when adding accessories to their machines. It will be evident that the Section can only cover the subject in the most general terms and so it is recommended that the owner, having decided that he wants to fit, for example, a luggage rack or carrier, seeks the advice of several local dealers and the owners of similar machines. This will give a good idea of what makes of carrier are easily available, and at what price. Talking to other owners will give some insight into the drawbacks or good points of any one make. A walk round the motorcycles in car parks or outside a dealer will often reveal the same sort of information.

The first priority when choosing accessories is to assess exactly what one needs. It is, for example, pointless to buy a large heavy-duty carrier which is designed to take the weight of fully laden panniers and topbox when all you need is a place to strap on a set of waterproofs and a lunchbox when going to work. Many accessory manufacturers have ranges of equipment to cater for the individual needs of different riders and this point should be borne in mind when looking through a dealer's catalogues. Having decided exactly what is required and the use to which the accessories are going to be put, the owner will need a few hints on what to look for when making the final choice. To this end the Section is now sub-divided to cover the more popular accessories fitted. Note that it is in no way a customizing guide, but merely seeks to outline the practical considerations to be taken into account when adding aftermarket equipment to a motorcycle.

Fairings and windscreens

A fairing is possibly the single, most expensive, aftermarket item to be fitted to any motorcycle and, therefore, requires the most thought before purchase. Fairings can be divided into two main groups: front fork mounted handlebar fairings and windscreens, and frame mounted fairings.

The first group, the front fork mounted fairings, are becoming far more popular than was once the case, as they offer several advantages over the second group. Front fork mounted fairings generally are much easier and quicker to fit, involve less modification to the motorcycle, do not as a rule restrict the steering lock, permit a wider selection of handlebar styles to be used, and offer adequate protection for much less money than the frame mounted type. They are also lighter, can be swapped easily between different motorcycles, and are available in a much greater variety of styles. Their main disadvantages are that they do not offer as much weather protection as the frame mounted types, rarely offer any storage space, and, if poorly fitted or naturally incompatible, can have an adverse effect on the stability of the motorcycle.

The second group, the frame mounted fairings, are secured so rigidly to the main frame of the motorcycle that they can offer a substantial amount of protection to motorcycle and rider in the event of a crash. They offer almost complete protection from the weather and, if double-skinned in construction, can provide a great deal of useful storage space. The feeling of peace, quiet and complete relaxation encountered when riding behind a good full fairing has to be experienced to be believed. For this reason full fairings are considered essential by most touring motorcyclists and by many people who ride all year round. The main disadvantages of this type are that fitting can take a long time, often involving removal or modification of standard motorcycle components, they restrict the steering lock and they can add up to about 40 lb to the weight of the machine. They do not usually affect the stability of the machine to any great extent once the front tyre pressure and suspension have been adjusted to compensate for the extra weight, but can be affected by sidewinds.

The first thing to look for when purchasing a fairing is the quality of the fittings. A good fairing will have strong, substantial brackets constructed from heavy-gauge tubing; the brackets must be shaped to fit the frame or forks evenly so that the minimum of stress is imposed on the assembly when it is bolted down. The brackets should be properly painted or finished – a nylon coating being the favourite of the better manufacturers – the nuts and bolts provided should be of the same thread and size standard as is used on the motorcycle and be properly plated. Look also for shakeproof locking nuts or locking washers to ensure that everything remains securely tightened down. The fairing shell is generally made from one of two materials: fibreglass or ABS plastic. Both have their advantages and disadvantages, but the main consideration for the owner is that fibreglass is much easier to repair in the event of damage occurring to the fairing. Whichever material is used, check that it is properly finished inside as well as out, that the edges are protected by beading and that the fairing shell is insulated from vibration by the use of rubber grommets at all mounting points. Also be careful to check that the windscreen is retained by plastic bolts which will snap on impact so that the windscreen will break away and not cause personal injury in the event of an accident.

Having purchased your fairing or windscreen, read the manufacturer's fitting instructions very carefully and check that you have all the necessary brackets and fittings. Ensure that the mounting brackets are located correctly and bolted down securely. Note that some manufacturers use hose clamps to retain the mounting brackets; these should be discarded as they are convenient to use but not strong enough for the task. Stronger clamps should be substituted; car exhaust pipe clamps of suitable size would be a good alternative. Ensure that the front forks can turn through the full steering lock available without fouling the fairing. With many types of frame-mounted fairing the handlebars will have to be altered or a different type fitted and the steering lock will be restricted by stops provided with the fittings. Also check that the fairing does not foul the front wheel or mudguard, in any steering position, under full fork compression. Re-route any cables, brake pipes or electrical wiring which may snag on the fairing and take great care to protect all electrical connections, using insulating tape. If the manufacturer's instructions are followed carefully at every stage no serious problems should be encountered. Remember that hydraulic pipes that have been disconnected must be carefully re-tightened and the hydraulic system purged of air bubbles by bleeding.

Two things will become immediately apparent when taking a motorcycle on the road for the first time with a fairing – the first is the tendency to underestimate the road speed because of the lack of wind pressure on the body. This must be very carefully watched until one has grown accustomed to riding behind the fairing. The second thing is the alarming increase in engine noise which is an unfortunate but inevitable by-product of fitting any type of fairing or windscreen, and is caused by normal engine noise being reflected, and in some cases amplified, by the flat surface of the fairing.

Luggage racks or carriers

Carriers are possibly the commonest item to be fitted to modern motorcycles. They vary enormously in size, carrying capacity, and durability. When selecting a carrier, always look for one which is made specifically for your machine and which is bolted on with as few separate brackets as possible. The universal-type carrier, with its mass of brackets and adaptor pieces, will generally prove too weak to be of any real use. A good carrier should bolt to the main frame, generally using the two suspension unit top mountings and a mudguard mounting bolt as attachment points, and have its luggage platform as low and as far forward as possible to minimise the effect of any load on the machine's stability. Look for good quality, heavy gauge tubing, good welding and good finish. Also ensure that the carrier does not prevent opening of the seat, sidepanels or tail compartment, as appropriate. When using a carrier, be very careful not to overload it. Excessive weight placed so high and so far to the rear of any motorcycle will have an adverse effect on the machine's steering and stability.

Luggage

Motorcycle luggage can be grouped under two headings: soft and hard. Both types are available in many sizes and styles and have advantages and disadvantages in use.

Soft luggage is now becoming very popular because of its lower cost and its versatility. Whether in the form of tankbags, panniers, or strap-on bags, soft luggage requires in general no brackets and no modification to the motorcycle. Equipment can be swapped easily from one motorcycle to another and can be fitted and removed in seconds. Awkwardly shaped loads can easily be carried. The disadvantages of soft luggage are that the contents cannot be secure against the casual thief, very little protection is afforded in the event of a crash, and waterproofing is generally poor. Also, in the case of panniers, carrying capacity is restricted to approximately 10 lb, although this amount will vary considerably depending on the manufacturer's recommendation. When purchasing soft luggage, look for good quality material, generally vinyl or nylon, with strong, well-stitched attachment points. It is always useful to have separate pockets, especially on tank bags, for items which will be needed on the journey. When purchasing a tank bag, look for one which

has a separate, well-padded, base. This will protect the tank's paintwork and permit easy access to the filler cap at petrol stations.

Hard luggage is confined to two types: panniers, and top boxes or tail trunks. Most hard luggage manufacturers produce matching sets of these items, the basis of which is generally that manufacturer's own heavy-duty luggage rack. Variations on this theme occur in the form of separate frames for the better quality panniers, fixed or quickly-detachable luggage, and in size and carrying capacity. Hard luggage offers a reasonable degree of security against theft and good protection against weather and accident damage. Carrying capacity is greater than that of soft luggage, around 15 – 20 lb in the case of panniers, although top boxes should never be loaded as much as their apparent capacity might imply. A top box should only be used for lightweight items, because one that is heavily laden can have a serious effect on the stability of the machine. When purchasing hard luggage look for the same good points as mentioned under fairings and windscreens, ie good quality mounting brackets and fittings, and well-finished fibreglass or ABS plastic cases. Again as with fairings, always purchase luggage made specifically for your motorcycle, using as few separate brackets as possible, to ensure that everything remains securely bolted in place. When fitting hard luggage, be careful to check that the rear suspension and brake operation will not be impaired in any way and remember that many pannier kits require re-siting of the indicators. Remember also that a non-standard exhaust system may make fitting extremely difficult.

Handlebars

The occupation of fitting alternative types of handlebar is extremely popular with modern motorcyclists, whose motives may vary from the purely practical, wishing to improve the comfort of their machines, to the purely aesthetic, where form is more important than function. Whatever the reason, there are several considerations to be borne in mind when changing the handlebars of your machine. If fitting lower bars, check carefully that the switches and cables do not foul the petrol tank on full lock and that the surplus length of cable, brake pipe, and electrical wiring are smoothly and tidily disposed of. Avoid tight kinks in cable or brake pipes which will produce stiff controls or the premature and disastrous failure of an overstressed component. If necessary, remove the petrol tank and re-route the cable from the engine/gearbox unit upwards, ensuring smooth gentle curves are produced. In extreme cases, it will be necessary to purchase a shorter brake pipe to overcome this problem. In the case of higher handlebars than standard it will almost certainly be necessary to purchase extended cables and brake pipes. Fortunately, many standard motorcycles have a custom version which will be equipped with higher handlebars and, therefore, factory-built extended components will be available from your local dealer. It is not usually necessary to extend electrical wiring, as switch clusters may be used on several different motorcycles, some being custom versions. This point should be borne in mind however when fitting extremely high or wide handlebars.

When fitting different types of handlebar, ensure that the mounting clamps are correctly tightened to the manufacturer's specifications and that cables and wiring, as previously mentioned, have smooth easy runs and do not snag on any part of the motorcycle throughout the full steering lock. Ensure that the fluid level in the front brake master cylinder remains level to avoid any chance of air entering the hydraulic system. Also check that the cables are adjusted correctly and that all handlebar controls operate correctly and can be easily reached when riding.

Crashbars

Crashbars, also known as engine protector bars, engine guards, or case savers, are extremely useful items of equipment which can contribute protection to the machine's structure if a crash occurs. They do not, as has been inferred in the US, prevent the rider from crashing, or necessarily prevent rider injury should a crash occur.

It is recommended that only the smaller, neater, engine protector type of crashbar is considered. This type will offer protection while restricting, as little as is possible, access to the engine and the machine's ground clearance. The crashbars should be designed for use specifically on your machine, and should be constructed of heavy-gauge tubing with strong, integral mounting brackets. Where possible, they should bolt to a strong lug on the frame, usually at the engine mounting bolts.

The alternative type of crashbar is the larger cage type. This type is not recommended in spite of their appearance which promises some protection to the rider as well as to the machine. The larger amount of leverage imposed by the size of this type of crashbar increases the risk of severe frame damage in the event of an accident. This type also decreases the machine's ground clearance and restricts access to the engine. The amount of protection afforded the rider is open to some doubt as the design is based on the premise that the rider will stay in the normally seated position during an accident, and the crash bar structure will not itself fail. Neither result can in any way be guaranteed.

As a general rule, always purchase the best, ie usually the most expensive, set of crashbars you an afford. The investment will be repaid by minimising the amount of damage incurred, should the machine be involved in an accident. Finally, avoid the universal type of crashbar. This should be regarded only as a last resort to be used if no alternative exists. With its usual multitude of separate brackets and spacers, the universal crashbar is far too weak in design and construction to be of any practical value.

Exhaust systems

The fitting of aftermarket exhaust systems is another extremely popular pastime amongst motorcyclists. The usual motive is to gain more performance from the engine but other considerations are to gain more ground clearance, to lose weight from the motorcycle, to obtain a more distinctive exhaust note or to find a cheaper alternative to the manufacturer's original equipment exhaust system. Original equipment exhaust systems often cost more and may well have a relatively short life. It should be noted that it is rare for an aftermarket exhaust system alone to give a noticeable increase in the engine's power output. Modern motorcycles are designed to give the highest power output possible allowing for factors such as quietness, fuel economy, spread of power, and long-term reliability. If there were a magic formula which allowed the exhaust system to produce more power without affecting these other considerations you can be sure that the manufacturers, with their large research and development facilities, would have found it and made use of it. Performance increases of a worthwhile and noticeable nature only come from well-tried and properly matched modifications to the entire engine, from the air filter, through the carburettors, port timing or camshaft and valve design, combustion chamber shape, compression ratio, and the exhaust system. Such modifications are well outside the scope of this manual but interested owners might refer to specialist books produced by the publisher of this manual which go into the subject in great detail.

Whatever your motive for wishing to fit an alternative exhaust system, be sure to seek expert advice before doing so. Changes to the carburettor jetting will almost certainly be required for which you must consult the exhaust system manufacturer. If he cannot supply adequately specific information it is reasonable to assume that insufficient development work has been carried out, and that particular make should be avoided. Other factors to be borne in mind are whether the exhaust system allows the use of both centre and side stands, whether it allows sufficient access to permit oil and filter changing and whether modifications are necessary to the standard exhaust system. Many two-stroke expansion chamber systems require the use of the standard exhaust pipe; this is all very well if the standard exhaust pipe and silencer are separate units but can cause problems if the two, as with so many modern two-strokes, are a one-piece unit. While the exhaust pipe can be removed easily by means of a hacksaw it is not so easy to refit the original silencer should you at any time wish to return the machine to standard trim. The same applies to several four-stroke systems.

On the subject of the finish of aftermarket exhausts, avoid black-painted systems unless you enjoy painting. As any trail-bike owner will tell you, rust has a great affinity for black exhausts and re-painting or rust removal becomes a task which must be carried out with monotonous regularity. A bright chrome finish is, as a general rule, a far better proposition as it is much easier to keep clean and to prevent rusting. Although the general finish of aftermarket exhaust systems is not always up to the standard of the original equipment the lower cost of such systems does at least reflect this fact.

When fitting an alternative system always purchase a full set of new exhaust gaskets, to prevent leaks. Fit the exhaust first to the cylinder head or barrel, as appropriate, tightening the retaining nuts or bolts by hand only and then line up the exhaust rear mountings. If the new system is a one-piece unit and the rear mountings do not line up exactly, spacers must be fabricated to take up the difference. Do not force the system into place as the stress thus imposed will rapidly cause cracks and splits to appear. Once all the mountings are loosely fixed, tighten the retaining nuts or bolts securely, being careful not to overtighten them. Where the motorcycle manufacturer's torque settings are available, these should be used. Do not forget to carry out any carburation changes recommended by the exhaust system's manufacturer.

Electrical equipment

The vast range of electrical equipment available to motorcyclists is so large and so diverse that only the most general outline can be given here. Electrical accessories vary from electronic ignition kits fitted to replace contact breaker points, to additional lighting at the front and rear, more powerful horns, various instruments and gauges, clocks, anti-theft systems, heated clothing, CB radios, radio-cassette players, and intercom systems, to name but a few of the more popular items of equipment.

As will be evident, it would require a separate manual to cover this subject alone and this section is therefore restricted to outlining a few basic rules which must be borne in mind when fitting electrical equipment. The first consideration is whether your machine's electrical system has enough reserve capacity to cope with the added demand of the accessories you wish to fit. The motorcycle's manufacturer or importer should be able to furnish this sort of information and may also be able to offer advice on uprating the electrical system. Failing this, a good dealer or the accessory manufacturer may be able to help. In some cases, more powerful generator components may be available, perhaps from another motorcycle in the manufacturer's range. The second consideration is the legal requirements in force in your area. The local police may be prepared to help with this point. In the UK for example, there are strict regulations governing the position and use of auxiliary riding lamps and fog lamps.

When fitting electrical equipment always disconnect the battery first to prevent the risk of a short-circuit, and be careful to ensure that all connections are properly made and that they are waterproof. Remember that many electrical accessories are designed primarily for use in cars and that they cannot easily withstand the exposure to vibration and to the weather. Delicate components must be rubber-mounted to insulate them from vibration, and sealed carefully to prevent the entry of rainwater and dirt. Be careful to follow exactly the accessory manufacturer's instructions in conjunction with the wiring diagram at the back of this manual.

Accessories – general

Accessories fitted to your motorcycle will rapidly deteriorate if not cared for. Regular washing and polishing will maintain the finish and will provide an opportunity to check that all mounting bolts and nuts are securely fastened. Any signs of chafing or wear should be watched for, and the cause cured as soon as possible before serious damage occurs.

As a general rule, do not expect the re-sale value of your motorcycle to increase by an amount proportional to the amount of money and effort put into fitting accessories. It is usually the case that an absolutely standard motorcycle will sell more easily at a better price than one that has been modified. If you are in the habit of exchanging your machine for another at frequent intervals, this factor should be borne in mind to avoid loss of money.

Fault diagnosis

Contents

1 Introduction

This Section provides an easy reference-guide to the more common faults that are likely to afflict your machine. Obviously, the opportunities are almost limitless for faults to occur as a result of obscure failures, and to try and cover all eventualities would require a book. Indeed, a number have been written on the subject.

Successful fault diagnosis is not a mysterious 'black art' but the application of a bit of knowledge combined with a systematic and logical approach to the problem. Approach any fault diagnosis by first accurately identifying the symptom and then checking through the list of possible causes, starting with the simplest or most obvious and progressing in stages to the most complex. Take nothing for granted, but above all apply liberal quantities of common sense.

The main symptom of a fault is given in the text as a major heading below which are listed, as Section headings, the various systems or areas which may contain the fault. Details of each possible cause for a fault and the remedial action to be taken are given, in brief, in the paragraphs below each Section heading. Further information should be sought in the relevant Chapter.

Engine does not start when turned over

2 No fuel flow to carburettor

● Fuel tank empty or level too low. Check that the tap is turned to 'On' or 'Reserve' position as required. If in doubt, prise off the fuel feed pipe at the carburettor end and check that fuel runs from pipe when the tap is turned on.
● Tank filler cap vent obstructed. This can prevent fuel from flowing into the carburettor float bowl bcause air cannot enter the fuel tank to replace it. The problem is more likely to appear when the machine is being ridden. Check by listening close to the filler cap and releasing it. A hissing noise indicates that a blockage is present. Remove the cap and clear the vent hole with wire or by using an air line from the inside of the cap.
● Fuel tap or filter blocked. Blockage may be due to accumulation of rust or paint flakes from the tank's inner surface or of foreign matter from contaminated fuel. Remove the tap and clean it and the filter. Look also for water droplets in the fuel.
● Fuel line blocked. Blockage of the fuel line is more likely to result from a kink in the line rather than the accumulation of debris.

3 Fuel not reaching cylinder

● Float chamber not filling. Caused by float needle or floats sticking in up position. This may occur after the machine has been left standing for an extended length of time allowing the fuel to evaporate. When this occurs a gummy residue is often left which hardens to a varnish-like substance. This condition may be worsened by corrosion and crystalline deposits produced prior to the total evaporation of contaminated fuel. Sticking of the float needle may also be caused by wear. In any case removal of the float chamber will be necessary for inspection and cleaning.
● Blockage in starting circuit, slow running circuit or jets. Blockage of these items may be attributable to debris from the fuel tank by-passing the filter system or to gumming up as described in paragraph 1. Water droplets in the fuel will also block jets and passages. The carburettor should be dismantled for cleaning.
● Fuel level too low. The fuel level in the float chamber is controlled by float height. The fuel level may increase with wear or damage but will never reduce, thus a low fuel level is an inherent rather than developing condition. Check the float height, renewing the float or needle if required.

4 Engine flooding

● Float valve needle worn or stuck open. A piece of rust or other debris can prevent correct seating of the needle against the valve seat thereby permitting an uncontrolled flow of fuel. Similarly, a worn needle or needle seat will prevent valve closure. Dismantle the carburettor float bowl for cleaning and, if necessary, renewal of the worn components.
● Fuel level too high. The fuel level is controlled by the float height which may increase due to wear of the float needle, pivot pin or operating tang. Check the float height, and make any necessary adjustments. A leaking float will cause an increase in fuel level, and thus should be renewed.
● Cold starting mechanism. Check the choke (starter mechanism) for correct operation. If the mechanism jams in the 'On' position subsequent starting of a hot engine will be difficult.
● Blocked air filter. A badly restricted air filter will cause flooding. Check the filter and clean or renew as required. A collapsed inlet hose will have a similar effect. Check that the air filter inlet has not become blocked by a rag or similar item.

5 No spark at plug

● Ignition switch not on.
● Engine stop switch off.
● Spark plug dirty, oiled or 'whiskered'. Because the induction mixture of a two-stroke engine is inclined to be of a rather oily nature it is comparatively easy to foul the plug electrodes, especially where there have been repeated attempts to start the engine. A machine used for short journeys will be more prone to fouling because the engine may never reach full operating temperature, and the deposits will not burn off. On rare occasions a change of plug grade may be required but the advice of a dealer should be sought before making such a change. On all two-stroke machines it is a sound precaution to carry a new spare spark plug for substitution in the event of fouling problems.
● Spark plug failure. Clean the spark plug thoroughly and reset the electrode gap. Refer to the spark plug section and the colour condition guide in Routine Maintenance. If the spark plug shorts internally or has sustained visible damage to the electrodes, core or ceramic insulator it should be renewed. On rare occasions a plug that appears to spark vigorously will fail to do so when refitted to the engine and subjected to the compression pressure in the cylinder.
● Spark plug cap or high tension (HT) lead faulty. Check condition and security. Replace if deterioration is evident. On rare occasions the cap may break down, thus preventing sparking. If this is suspected, fit a new cap as a precaution.
● Spark plug cap loose. Check that the spark plug cap fits securely over the plug and, where fitted, the screwed terminal on the plug end is secure.
● Shorting due to moisture. Certain parts of the ignition system are susceptible to shorting when the machine is ridden or parked in wet weather. Check particularly the area from the spark plug cap back to the ignition coil. A water dispersant spray may be used to dry out waterlogged components. Recurrence of the problem can be prevented by using an ignition sealant spray after drying out and cleaning.
● Ignition or stop switch shorted. May be caused by water corrosion or wear. Water dispersant and contact cleaning sprays may be used. If this fails to overcome the problem dismantling and visual inspection of the switches will be required.
● Shorting or open circuit in wiring. Failure in any wire connecting any of the ignition components will cause ignition malfunction. Check also that all connections are clean, dry and tight.
● Ignition coil failure. Check the coil, referring to Chapter 4.

6 Weak spark at plug

● Feeble sparking at the plug may be caused by any of the faults mentioned in the preceding Section other than those items in the first three paragraphs. Check first the spark plug, this being the most likely culprit.

7 Compression low

● Spark plug loose. This will be self-evident on inspection, and may be accompanied by a hissing noise when the engine is turned over.

Remove the plug and check that the threads in the cylinder head are not damaged. Check also that the plug sealing washer is in good condition.
● Cylinder head gasket leaking. This condition is often accompanied by loss of coolant through overheating, and may be caused by insufficiently tightened cylinder head fasteners, a warped cylinder head or mechanical failure of the gasket material. Re-torqueing the fasteners to the correct specification may seal the leak in some instances but if damage has occurred this course of action will provide, at best, only a temporary cure.
● Low crankcase compression. This can be caused by worn main bearings and seals and will upset the incoming fuel/air mixture. A good seal in these areas is essential on any two-stroke engine.
● Piston rings sticking or broken. Sticking of the piston rings may be caused by seizure due to lack of lubrication or overheating as a result of poor carburation or incorrect fuel type. Gumming of the rings may result from lack of use, or carbon deposits in the ring grooves. Broken rings result from over-revving, over-heating or general wear. In either case a top-end overhaul will be required.

Engine stalls after starting

8 General causes

● Improper cold start mechanism operation. Check that the operating control functions smoothly. A cold engine may not require application of an enriched mixture to start initially but may baulk without choke once firing. Likewise a hot engine may start with an enriched mixture but will stop almost immediately if the choke is inadvertently in operation.
● Ignition malfunction. See Section 9. Weak spark at plug.
● Carburettor incorrectly adjusted. Maladjustment of the mixture strength or idle speed may cause the engine to stop immediately after starting. See Chapter 3.
● Fuel contamination. Check for filter blockage by debris or water which reduces, but does not completely stop, fuel flow, or blockage of the slow speed circuit in the carburettor by the same agents. If water is present it can often be seen as droplets in the bottom of the float bowl. Clean the filter and, where water is in evidence, drain and flush the fuel tank and float bowl.
● Intake air leak. Check for security of the carburettor mounting and hose connections, and for cracks or splits in the hoses. Check also that the carburettor top is secure and that the oil pump/inlet stub feed pipe union is tight.
● Air filter blocked or omitted. A blocked filter will cause an over-rich mixture; the omission of a filter will cause an excessively weak mixture. Both conditions will have a detrimental effect on carburation. Clean or renew the filter as necessary.
● Fuel filler cap air vent blocked. Usually caused by dirt or water. Clean the vent orifice.
● Choked exhaust system. Caused by excessive carbon build-up in the system, particularly around the silencer baffles. Refer to Routine Maintenance for further information.
● Excessive carbon build-up in the engine. This can result from failure to decarbonise the engine at the specified interval or through excessive oil consumption. Check pump adjustment.

Poor running at idle and low speed

9 Weak spark at plug or erratic firing

● Battery voltage low. In certain conditions low battery charge, especially when coupled with a badly sulphated battery, may result in misfiring. If the battery is in good general condition it should be recharged; an old battery suffering from sulphated plates should be renewed.
● Spark plug fouled, faulty or incorrectly adjusted. See Section 5 or refer to Routine Maintenance.
● Spark plug cap or high tension lead shorting. Check the condition of both these items ensuring that they are in good condition and dry and that the cap is fitted correctly.
● Spark plug type incorrect. Fit plug of correct type and heat range

as given in Specifications. In certain conditions a plug of hotter or colder type may be required for normal running.
● Ignition timing incorrect. Check the ignition timing ensuring that the retard is functioning correctly.
● Faulty ignition coil. Partial failure of the coil internal insulation will diminish the performance of the coil. No repair is possible, a new component must be fitted.
● Defective flywheel generator ignition source. Refer to Chapter 4 for further details on test procedures.

10 Fuel/air mixture incorrect

● Intake air leak. Check carburettor mounting and air cleaner hose for security and signs of splitting. Ensure that carburettor top is tight and that the oil pump/inlet stub feed pipe union is tight.
● Mixture strength incorrect. Adjust slow running mixture strength using pilot adjustment screw.
● Pilot jet or slow running circuit blocked. The carburettor should be removed and dismantled for thorough cleaning. Blow through all jets and air passages with compressed air to clear obstructions.
● Air cleaner clogged or omitted. Clean or fit air cleaner element as necessary. Check also that the element and air filter cover are correctly seated.
● Cold start mechanism in operation. Check that the choke has not been left on inadvertently and the operation is correct.
● Fuel level too high or too low. Check the float height, renewing float or needle if required. See Section 3 or 4.
● Fuel tank air vent obstructed. Obstructions usually caused by dirt or water. Clean vent orifice.

11 Compression low

● See Section 7.

Acceleration poor

12 General causes

● All items as for previous Section.
● Choked air filter. Failure to keep the air filter element clean will allow the build-up of dirt with proportional loss of performance. In extreme cases of neglect acceleration will suffer.
● Choked exhaust system. This can result from failure to remove accumulations of carbon from the silencer baffles at the prescribed intervals. The increased back pressure will make the machine noticeably sluggish. Refer to Routine Maintenance for further information on decarbonisation.
● Excessive carbon build-up in the engine. This can result from failure to decarbonise the engine at the specified interval or through excessive oil consumption. Check pump adjustment.
● Ignition timing incorrect. Check the ignition timing as described in Chapter 4. Since no provision for adjustment exists, test the electronic ignition components and renew as required.
● Carburation fault. See Section 10.
● Mechanical resistance. Check that the brakes are not binding. On small machines in particular note that the increased rolling resistance caused by under-inflated tyres may impede acceleration.
● Faulty ATAC components (MTX200 only). Refer to Chapter 3.

Poor running or lack of power at high speeds

13 Weak spark at plug or erratic firing

● All items as for Section 9.
● HT lead insulation failure. Insulation failure of the HT lead and

spark plug cap due to old age or damage can cause shorting when the engine is driven hard. This condition may be less noticeable, or not noticeable at all at lower engine speeds.
● Faulty ATAC mechanism (MTX200 only). If the valve is not closing correctly at the specified speed, power will be reduced but the fault may also produce a misfire. On 125 models, check that the cylinder head and barrel chamber covers are securely fitted and not leaking. Refer to Chapter 3.

14 Fuel/air mixture incorrect

● All items as for Section 10, with the exception of items relative exclusively to low speed running.
● Main jet blocked. Debris from contaminated fuel, or from the fuel tank, and water in the fuel can block the main jet. Clean the fuel filter, the float bowl area, and if water is present, flush and refill the fuel tank.

● Main jet is the wrong size. The standard carburettor jetting is for sea level atmospheric pressure. For high altitudes, usually above 5000 ft, a smaller main jet will be required.
● Jet needle and needle jet worn. These can be renewed individually but should be renewed as a pair. Renewal of both items requires partial dismantling of the carburettor.
● Air bleed holes blocked. Dismantle carburettor and use compressed air to blow out all air passages.
● Reduced fuel flow. A reduction in the maximum fuel flow from the fuel tank to the carburettor will cause fuel starvation, proportionate to the engine speed. Check for blockages through debris or a kinked fuel line.

15 Compression low

● See Section 7.

Knocking or pinking

16 General causes

● Carbon build-up in combustion chamber. After high mileages have been covered large accumulations of carbon may occur. This may glow red hot and cause premature ignition of the fuel/air mixture, in advance of normal firing by the spark plug. Cylinder head removal will be required to allow inspection and cleaning.
● Fuel incorrect. A low grade fuel, or one of poor quality may result in compression induced detonation of the fuel resulting in knocking and pinking noises. Old fuel can cause similar problems. A too highly leaded fuel will reduce detonation but will accelerate deposit formation in the combustion chamber and may lead to early pre-ignition as described in item 1.
● Spark plug heat range incorrect. Uncontrolled pre-ignition can result from the use of a spark plug the heat range of which is too hot.
● Weak mixture. Overheating of the engine due to a weak mixture can result in pre-ignition occurring where it would not occur when engine temperature was within normal limits. Maladjustment, blocked jets or passages and air leaks can cause this condition.

Overheating

17 Firing incorrect

● Spark plug fouled, defective or maladjusted. See Section 5.
● Spark plug type incorrect. Refer to the Specifications and ensure that the correct plug type is fitted.
● Incorrect ignition timing. Timing that is far too much advanced or far too much retarded will cause overheating. Check the ignition timing is correct.

18 Fuel/air mixture incorrect

● Slow speed mixture strength incorrect. Adjust pilot air screw.
● Main jet wrong size. The carburettor is jetted for sea level atmospheric conditions. For high altitudes, usually above 5000 ft, a smaller main jet will be required.
● Air filter badly fitted or omitted. Check that the filter element is in place and that it and the air filter box cover are sealing correctly. Any leaks will cause a weak mixture.
● Induction air leaks. Check the security of the carburettor mounting and hose connection, and for cracks and splits in the hoses. Check also that the carburettor top is secure and that the oil pump/inlet stub feed pipe union is tight.
● Fuel level too low. See Section 3.
● Fuel tank filler cap air vent obstructed. Clear blockage.

19 Lubrication inadequate

● Oil pump settings incorrect. The oil pump settings are of great importance s

ince the quantities of oil being injected are very small. Any variation in oil delivery will have a significant effect on the engine. Refer to Chapter 3 for further information.
● Oil tank empty or low. This will have disastrous consequences if left unnoticed. Check and replenish tank regularly.
● Transmission oil low or worn out. Check the level regularly and investigate any loss of oil. If the oil level drops with no sign of external leakage it is likely that the crankshaft main bearing oil seals are worn, allowing transmission oil to be drawn into the crankcase during induction.

20 Miscellaneous causes

● Faulty thermostat. If the thermostat fails in the closed position the engine will rapidly overheat. Renew the thermostat. See Chapter 2.
● Radiator fins clogged. Accumulated debris in the radiator core will gradually reduce its ability to dissipate heat generated by the engine. It is worth noting that during the summer months dead insects can cause as many problems in this respect as road dirt and mud during the winter. Cleaning is best carried out by dislodging the debris with a high pressure hose from the back of the radiator. Once cleaned it is worth painting the matrix with a heat-dispersant matt black paint both to assist cooling and to prevent external corrosion. The fitting of some sort of mesh guard will help prevent the fins from becoming clogged, but make sure that this does not itself prevent adequate cooling.

Clutch operating problems

21 Clutch slip

● No clutch lever play. Adjust clutch lever end play according to the procedure in Routine Maintenance.
● Friction plates worn or warped. Overhaul clutch assembly, replacing plates out of specification.
● Steel plates worn or warped. Overhaul clutch assembly, replacing plates out of specification.
● Clutch spring broken or worn. Old or heat-damaged (from slipping clutch) springs should be replaced with new ones.
● Clutch inner cable snagging. Caused by a frayed cable or kinked outer cable. Replace the cable with a new one. Repair of a frayed cable is not advised.
● Clutch release mechanism defective. Worn or damaged parts in the clutch release mechanism could include the pushrod actuating arm. Replace parts as necessary.
● Clutch centre and outer drum worn. Severe indentation by the clutch plate tangs of the channels in the centre and drum will cause snagging of the plates preventing correct engagement. If this damage occurs, renewal of the worn components is required.

● Lubricant incorrect. Use of a transmission lubricant other than that specified may allow the plates to slip.

22 Clutch drag

● Clutch lever play excessive.
● Clutch plates warped or damaged. This will cause a drag on the clutch, causing the machine to creep. Overhaul clutch assembly.
● Clutch spring tension uneven. Usually caused by a sagged or broken spring. Check and replace springs.
● Transmission oil deteriorated. Badly contaminated transmission oil and a heavy deposit of oil sludge on the plates will cause plate sticking. The oil recommended for this machine is of the detergent type, therefore it is unlikely that this problem will arise unless regular oil changes are neglected.
● Transmission oil viscosity too high. Drag in the plates will result from the use of an oil with too high a viscosity. In very cold weather clutch drag may occur until the engine has reached operating temperature.
● Clutch centre and outer drum worn. Indentation by the clutch plate tangs of the channels in the centre and drum will prevent easy plate disengagement. If the damage is light the affected areas may be dressed with a fine file. More pronounced damage will necessitate renewal of the components.
● Clutch outer drum seized to shaft. Lack of lubrication, severe wear or damage can cause the drum to seize to the shaft. Overhaul of the clutch, and perhaps the transmission, may be necessary to repair damage.
● Clutch release mechanism defective. Worn or damaged release mechanism parts can stick and fail to provide leverage. Overhaul clutch cover components.
● Loose clutch centre nut. Causes drum and centre misalignment, putting a drag on the engine. Engagement adjustment continually varies. Overhaul clutch assembly.

Gear selection problems

23 Gear lever does not return

● Weak or broken return spring. Renew the spring.
● Gearchange shaft bent or seized. Distortion of the gearchange shaft often occurs if the machine is dropped heavily on the gear lever. Provided that damage is not severe straightening of the shaft is permissible.

24 Gear selection difficult or impossible

● Clutch not disengaging fully. See Section 22.
● Gearchange shaft bent. This often occurs if the machine is dropped heavily on the gear lever. Straightening of the shaft is permissible if the damage is not too great.
● Gearchange claw arms, teeth or pins worn or damaged. Wear or breakage of any of these items may cause difficulty in selecting one or more gears. Overhaul the selector mechanism.
● Selector drum stopper cam or detent roller arm damaged. Failure, rather than wear of these items may jam the drum thereby preventing gearchanging or causing false selection at high speed.
● Selector forks bent or seized. This can be caused by dropping the machine heavily on the gearchange lever or as a result of lack of lubrication. Though rare, bending of a shaft can result from a missed gearchange or false selection at high speed.
● Selector fork claw end and guide pin wear. Pronounced wear of these items and the grooves in the selector drum can lead to imprecise selection and, eventually, no selection. Renewal of the worn components will be required.
● Structural failure. Failure of any one component of the mechanism will result in improper or fouled gear selection.

25 Jumping out of gear

● Detent roller arm assembly worn or damaged. Wear of the roller and the cam with which it locates and breakage of the detent spring can cause imprecise gear selection resulting in jumping out of gear. Renew the damaged components.
● Gear pinion dogs worn or damaged. Rounding off the dog edges and the mating recesses in adjacent pinion can lead to jumping out of gear when under load. The gears should be inspected and renewed. Attempting to reprofile the dogs is not recommended.
● Selector forks, drum and pinion grooves worn. Extreme wear of these interconnected items can occur after high mileages especially when lubrication has been neglected. The worn components must be renewed.
● Gear pinions, bushes and shafts worn. Renew the worn components.
● Bent gearchange shaft. Often caused by dropping the machine on the gear lever.
● Gear pinion tooth broken. Chipped teeth are unlikely to cause jumping out of gear once the gear has been selected fully; a tooth which is completely broken off, however, may cause problems in this respect and in any event will cause transmission noise.

26 Overselection

● Claw arm pressure spring weak or broken. Renew the spring.
● Stopper arm spring worn or broken. Renew the spring.
● Gearchange arm stop pads worn. Repairs can be made by welding and reprofiling with a file.
● Selector limiter claw components (where fitted) worn or damaged. Renew the damaged items.

Abnormal engine noise

27 Knocking or pinking

● See Section 16.

28 Piston slap or rattling from cylinder

● Cylinder bore/piston clearance excessive. Resulting from wear, or partial seizure. This condition can often be heard as a high, rapid tapping noise when the engine is under little or no load, particularly when power is just beginning to be applied. Reboring to the next correct oversize should be carried out and a new oversize piston fitted.
● Connecting rod bent. This can be caused by over-revving, trying to start a very badly flooded engine (resulting in a hydraulic lock in the cylinder) or by earlier mechanical failure. Attempts at straightening a bent connecting rod from a high performance engine are not recommended. Careful inspection of the crankshaft should be made before renewing the damaged connecting rod.
● Gudgeon pin, piston boss bore or small-end bearing wear or seizure. Excess clearance or partial seizure between normal moving parts of these items can cause continuous or intermittent tapping noises. Rapid wear or seizure is caused by lubrication starvation.
● Piston rings worn, broken or sticking. Renew the rings after careful inspection of the piston and bore.

29 Other noises

● Big-end bearing wear. A pronounced knock from within the crankcase which worsens rapidly is indicative of big-end bearing failure as a result of extreme normal wear or lubrication failure. Remedial action in the form of a bottom end overhaul should be taken; continuing to run the engine will lead to further damage including the possibility of connecting rod breakage.

● Main bearing failure. Extreme normal wear or failure of the main bearings is characteristically accompanied by a rumble from the crankcase and vibration felt through the frame and footrests. Renew the worn bearings and carry out a very careful examination of the crankshaft.

● Crankshaft excessively out of true. A bent crank may result from over-revving or damage from an upper cylinder component or gearbox failure. Damage can also result from dropping the machine on either crankshaft end. Straightening of the crankshaft is not be possible in normal circumstances; a replacement item should be fitted.

● Engine mounting loose. Tighten all the engine mounting nuts and bolts.

● Cylinder base gasket leaking. The noise most often associated with a leaking head gasket is a high pitched squeaking, although any other noise consistent with gas being forced out under pressure from a small orifice can also be emitted. Gasket leakage is often accompanied by oil seepage from around the mating joint or from the cylinder retaining nuts. Leakage results from insufficient or uneven tightening of the fasteners, or from random mechanical failure. Retightening to the correct torque figure will, at best, only provide a temporary cure. The gasket should be renewed at the earliest opportunity.

● Exhaust system leakage. Popping or crackling in the exhaust system, particularly when it occurs with the engine on the overrun, indicates a poor joint either at the cylinder port or at the exhaust pipe/silencer connection. Failure of the gasket or looseness of the clamp should be looked for.

● ATAC system leakage. While the components are only fitted to the MTX 200 model the passages are still present on 125 models and are sealed by separate covers which may work loose to produce similar symptoms to those described for the exhaust system.

Abnormal transmission noise

30 Clutch noise

● Clutch outer drum/friction plate tang clearance excessive.
● Clutch outer drum/spacer clearance excessive.
● Clutch outer drum/thrust washer clearance excessive.
● Primary drive gear teeth worn or damaged.
● Clutch shock absorber assembly worn or damaged.

31 Transmission noise

● Bearing or bushes worn or damaged. Renew the affected components.
● Gear pinions worn or chipped. Renew the gear pinions.
● Metal chips jammed in gear teeth.This can occur when pieces of metal from any failed component are picked up by a meshing pinion. The condition will lead to rapid bearing wear or early gear failure.
● Engine/transmission oil level too low. Top up immediately to prevent damage to gearbox and engine.
● Gearchange mechanism worn or damaged. Wear or failure of certain items in the selection and change components can induce mis-selection of gears (see Section 24) where incipient engagement of more than one gear set is promoted. Remedial action, by the overhaul of the gearbox, should be taken without delay.
● Chain snagging on cases or cycle parts. A badly worn chain or one that is excessively loose may snag or smack against adjacent components.

Exhaust smokes excessively

32 White/blue smoke (caused by oil burning)

● Cylinder cracked, worn or scored. These conditions may be caused by overheating, lack of lubrication, component failure or advanced normal wear. The cylinder barrel should be renewed and, if necessary, a new piston fitted.
● Oil pump settings incorrect. Check and reset the oil pump as described in Routine Maintenance.

● Crankshaft main bearing oil seals worn. Wear in the main bearing oil seals, often in conjunction with wear in the bearings themselves, can allow transmission oil to find its way into the crankcase and thence to the combustion chamber. This condition is often indicated by a mysterious drop in the transmission oil level with no sign of external leakage.

● Accumulated oil deposits in exhaust system. If the machine is used for short journeys only it is possible for the oil residue in the exhaust gases to condense in the relatively cool silencer. If the machine is then taken for a longer run in hot weather, the accumulated oil will burn off producing ominous smoke from the exhaust.

33 Black smoke (caused by over-rich mixture)

● Air filter element clogged. Clean or renew the element.
● Main jet loose or too large. Remove the float chamber to check for tightness of the jet. If the machine is used at high altitudes rejetting will be required to compensate for the lower atmospheric pressure.
● Cold start mechanism jammed on. Check that the mechanism works smoothly and correctly.
● Fuel level too high. The fuel level is controlled by the float height which can increase as a result of wear or damage. Remove the float bowl and check the float height. Check also that floats have not punctured; a punctured float will lose buoyancy and allow an increased fuel level.
● Float valve needle stuck open. Caused by dirt or a worn valve. Clean the float chamber or renew the needle and, if necessary, the valve seat.

Poor handling or roadholding

34 Directional instability

● Steering head bearing adjustment too tight. This will cause rolling or weaving at low speeds. Re-adjust the bearings.
● Steering head bearing worn or damaged. Correct adjustment of the bearing will prove impossible to achieve if wear or damage has occurred. Inconsistent handling will occur including rolling or weaving at low speed and poor directional control at indeterminate higher speeds. The steering head bearing should be dismantled for inspection and renewed if required. Lubrication should also be carried out.
● Bearing races pitted or dented. Impact damage caused, perhaps, by an accident or riding over a pot-hole can cause indentation of the bearing, usually in one position. This should be noted as notchiness when the handlebars are turned. Renew and lubricate the bearings.
● Steering stem bent. This will occur only if the machine is subjected to a high impact such as hitting a curb or a pot-hole. The bottom yoke should be renewed; do not attempt to straighten the stem.
● Front or rear tyre pressures too low.
● Front or rear tyre worn. General instability, high speed wobbles and skipping over white lines indicates that tyre renewal may be required. Tyre induced problems, in some machine/tyre combinations, can occur even when the tyre in question is by no means fully worn.
● Swinging arm bearings worn. Difficulties in holding line, particularly when cornering or when changing power settings indicates wear in the swinging arm bearings. The swinging arm should be removed from the machine and the bearings renewed.
● Swinging arm flexing. The symptoms given in the preceding paragraph will also occur if the swinging arm fork flexes badly. This can be caused by structural weakness as a result of corrosion, fatigue or impact damage, or because the rear wheel spindle is slack.
● Wheel bearings worn. Renew the worn bearings.
● Loose wheel spokes. The spokes should be tightened evenly to maintain tension and trueness of the rim.
● Tyres unsuitable for machine. Not all available tyres will suit the characteristics of the frame and suspension, indeed, some tyres or tyre combinations may cause a transformation in the handling characteristics. If handling problems occur immediately after changing to a new tyre type or make, revert to the original tyres to see whether an improvement can be noted. In some instances a change to what are, in fact, suitable tyres may give rise to handling deficiences. In this case a thorough check should be made of all frame and suspension items which affect stability.

35 Steering bias to left or right

● Rear wheel out of alignment. Caused by uneven adjustment of chain tensioner adjusters allowing the wheel to be askew in the fork ends. A bent rear wheel spindle will also misalign the wheel in the swinging arm.
● Wheels out of alignment. This can be caused by impact damage to the frame, swinging arm, wheel spindles or front forks. Although occasionally a result of material failure or corrosion it is usually as a result of a crash.
● Front forks twisted in the steering yokes. A light impact, for instance with a pot-hole or low curb, can twist the fork legs in the steering yokes without causing structural damage to the fork legs or the yokes themselves. Re-alignment can be made by loosening the yoke pinch bolts, wheel spindle and mudguard bolts. Re-align the wheel with the handlebars and tighten the bolts working upwards from the wheel spindle. This action should be carried out only when there is no chance that structural damage has occurred.

36 Handlebar vibrates or oscillates

● Tyres worn or out of balance. Either condition, particularly in the front tyre, will promote shaking of the fork assembly and thus the handlebars. A sudden onset of shaking can result if a balance weight is displaced during use.
● Tyres badly positioned on the wheel rims. A moulded line on each wall of a tyre is provided to allow visual verification that the tyre is correctly positioned on the rim. A check can be made by rotating the tyre; any misalignment will be immediately obvious.
● Wheel rims warped or damaged. Inspect the wheels for runout as described in Routine Maintenance.
● Swinging arm bearings worn. Renew the bearings.
● Wheel bearings worn. Renew the bearings.
● Steering head bearings incorrectly adjusted. Vibration is more likely to result from bearings which are too loose rather than too tight. Re-adjust the bearings.
● Loosen fork component fasteners. Loose nuts and bolts holding the fork legs, wheel spindle, mudguards or steering stem can promote shaking at the handlebars. Fasteners on running gear such as the forks and suspension should be check tightened occasionally to prevent dangerous looseness of components occurring.
● Engine mounting bolts loose. Tighten all fasteners.

37 Poor front fork performance

● Damping fluid level incorrect. If the fluid level is too low poor suspension control will occur resulting in a general impairment of roadholding and early loss of tyre adhesion when cornering and braking. Too much oil is unlikely to change the fork characteristics unless severe overfilling occurs when the fork action will become stiffer and oil seal failure may occur.
● Damping oil viscosity incorrect. The damping action of the fork is directly related to the viscosity of the damping oil. The lighter the oil used, the less will be the damping action imparted. For general use, use the recommended viscosity of oil, changing to a slightly higher or heavier oil only when a change in damping characteristic is required. Overworked oil, or oil contaminated with water which has found its way past the seals, should be renewed to restore the correct damping performance and to prevent bottoming of the forks.
● Damping components worn or corroded. Advanced normal wear of the fork internals is unlikely to occur until a very high mileage has been covered. Continual use of the machine with damaged oil seals which allows the ingress of water, or neglect, will lead to rapid corrosion and wear. Dismantle the forks for inspection and overhaul.
● Weak fork springs. Progressive fatigue of the fork springs, resulting in a reduced spring free length, will occur after extensive use. This condition will promote excessive fork dive under braking, and in its advanced form will reduce the at-rest extended length of the forks and thus the fork geometry. Renewal of the springs as a pair is the only satisfactory course of action.
● Bent stanchions or corroded stanchions. Both conditions will prevent correct telescoping of the fork legs, and in an advanced state

can cause sticking of the fork in one position. In a mild form corrosion will cause stiction of the fork thereby increasing the time the suspension takes to react to an uneven road surface. Bent fork stanchions should be attended to immediately because they indicate that impact damage has occurred, and there is a danger that the forks will fail with disastrous consequences.
● Incorrect, or more likely, unequal fork leg air pressures.

38 Front fork judder when braking (see also Section 50)

● Wear between the fork stanchions and the fork legs. Renewal of the affected components is required.
● Slack steering head bearings. Re-adjust the bearings.
● Warped brake disc or drum. If irregular braking action occurs fork judder can be induced in what are normally serviceable forks. Renew the damaged brake components.

39 Poor rear suspension performance

● Rear suspension unit damper worn out or leaking. The damping performance of most rear suspension units falls off with age. This is a gradual process, and thus may not be immediately obvious. Indications of poor damping include hopping of the rear end when cornering or braking, and a general loss of positive stability.
● Weak rear spring. If the suspension unit spring fatigues it will promote excessive pitching of the machine and reduce the ground clearance when cornering. Although replacement springs are available separately from the rear suspension damper unit it is probable that if spring fatigue has occurred the damper unit will also require renewal.
● Linkage bearings worn or improperly lubricated. Renew any damaged components.
● Swinging arm flexing or bearings worn. See Sections 34 and 36.
● Bent suspension unit damper rod. This is likely to occur only if the machine is dropped or if seizure of the piston occurs. Renew the unit.

Abnormal frame and suspension noise

40 Front end noise

● Oil level low or too thin. This can cause a 'spurting' sound and is usually accompanied by irregular fork action.
● Spring weak or broken. Makes a clicking or scraping sound. Fork oil will have a lot of metal particles in it.
● Steering head bearings loose or damaged. Clicks when braking. Check, adjust or replace.
● Fork clamps loose. Make sure all fork clamp pinch bolts are tight.
● Fork stanchion bent. Good possibility if machine has been dropped. Repair or replace tube.

41 Rear suspension noise

● Fluid level too low. Leakage of a suspension unit, usually evident by oil on the outer surfaces, can cause a spurting noise. The suspension unit should be renewed.
● Defective rear suspension unit with internal damage. Renew the suspension unit.
● Linkage bearings worn or unlubricated. Referring to Chapter 5, dismantle the rear suspension, renew any damaged components and reassemble, lubricating all bearings.

Brake problems

42 Brakes are spongy or ineffective – disc brakes

● Air in brake circuit. This is only likely to happen in service due to neglect in checking the fluid level or because a leak has developed. The problem should be identified and the brake system bled of air.

● Pad worn. Check the pad wear against the wear indicators provided and renew the pads if necessary.
● Contaminated pads. Cleaning pads which have been contaminated with oil, grease or brake fluid is unlikely to prove successful; the pads should be renewed.
● Pads glazed. This is usually caused by overheating. The surface of the pads may be roughened using glass-paper or a fine file.
● Brake fluid deterioration. A brake which on initial operation is firm but rapidly becomes spongy in use may be failing due to water contamination of the fluid. The fluid should be drained and then the system refilled and bled.
● Master cylinder seal failure. Wear or damage of master cylinder internal parts will prevent pressurisation of the brake fluid. Overhaul the master cylinder unit.
● Caliper seal failure. This will almost certainly be obvious by loss of fluid, a lowering of fluid in the master cylinder reservoir and contamination of the brake pads and caliper. Overhaul the caliper assembly.

43 Brakes drag – disc brakes

● Disc warped. The disc must be renewed.
● Caliper piston, caliper or pads corroded. The brake caliper assembly is vulnerable to corrosion due to water and dirt, and unless cleaned at regular intervals and lubricated in the recommended manner, will become sticky in operation.
● Piston seal deteriorated. The seal is designed to return the piston in the caliper to the retracted position when the brake is released. Wear or old age can affect this function. The caliper should be overhauled if this occurs.
● Brake pad damaged. Pad material separating from the backing plate due to wear or faulty manufacture. Renew the pads. Faulty installation of a pad also will cause dragging.
● Wheel spindle bent. The spindle may be straightened if no structural damage has occurred.
● Brake lever not returning. Check that the lever works smoothly throughout its operating range and does not snag on any adjacent cycle parts. Lubricate the pivot if necessary.
● Twisted caliper mounting bracket. This is likely to occur only after impact in an accident. No attempt should be made to re-align the caliper; the bracket should be renewed.

44 Brake lever or pedal pulsates in operation – disc brakes

● Disc warped or irregularly worn. The disc must be renewed.
● Wheel spindle bent. The spindle may be straightened provided no structural damage has occurred.

45 Disc brake noise

● Brake squeal. This can be caused by dust on the pads, usually in combination with glazed pads, or other contamination from oil, grease, brake fluid or corrosion. Persistent squealing which cannot be traced to any of the normal causes can often be cured by applying a thin layer of high temperature silicone grease to the rear of the pads. Make absolutely certain that no grease is allowed to contaminate the braking surface of the pads.
● Glazed pads. This is usually caused by high temperatures or contamination. The pad surfaces may be roughened using glass-paper or a fine file. If this approach does not effect a cure the pads should be renewed.
● Disc warped. This can cause a chattering, clicking or intermittent squeal and is usually accompanied by a pulsating brake lever or uneven braking. The disc must be renewed.
● Brake pads fitted incorrectly or undersize. Longitudinal play in the pads due to omission of the locating springs or because pads of the wrong size have been fitted will cause a single tapping noise every time the brake is operated. Inspect the pads for correct installation and security.

46 Brakes are spongy or ineffective – drum brakes

● Brake cable deterioration. Damage to the outer cable by stretching or being trapped will give a spongy feel to the brake lever. The cable should be renewed. A cable which has become corroded due to old age or neglect of lubrication will partially seize making operation very heavy. Lubrication at this stage may overcome the problem but the fitting of a new cable is recommended.
● Worn brake linings. Determine lining wear using the external brake wear indicator on the brake backplate, or by removing the wheel and withdrawing the brake backplate. Renew the shoe/lining units as a pair if the linings are worn below the recommended limit.
● Worn brake camshaft. Wear between the camshaft and the bearing surface will reduce brake feel and reduce operating efficiency. Renewal of one or both items will be required to rectify the fault.
● Worn brake cam and shoe ends. Renew the worn components.
● Linings contaminated with dust or grease. Any accumulations of dust should be cleaned from the brake assembly and drum using a petrol dampened cloth. Do not blow or brush off the dust because it is asbestos based and thus harmful if inhaled. Light contamination from grease can be removed from the surface of the brake linings using a solvent; attempts at removing heavier contamination are less likely to be successful because some of the lubricant will have been absorbed by the lining material which will severely reduce the braking performance.

47 Brake drag – drum brakes

● Incorrect adjustment. Re-adjust the brake operating mechanism.
● Drum warped or oval. This can result from overheating or impact or uneven tension of the wheel spokes. The condition is difficult to correct, although if slight ovality only occurs, skimming the surface of the brake drum can provide a cure. This is work for a specialist engineer. Renewal of the complete wheel hub is normally the only satisfactory solution.
● Weak brake shoe return springs. This will prevent the brake lining/shoe units from pulling away from the drum surface once the brake is released. The springs should be renewed.
● Brake camshaft, lever pivot or cable poorly lubricated. Failure to attend to regular lubrication of these areas will increase operating resistance which, when compounded, may cause tardy operation and poor release movement.

48 Brake lever or pedal pulsates in operation – drum brakes

● Drums warped or oval. This can result from overheating or impact or uneven spoke tension. This condition is difficult to correct, although if slight ovality only occurs skimming the surface of the drum can provide a cure. This is work for a specialist engineer. Renewal of the hub is normally the only satisfactory solution.

49 Drum brake noise

● Drum warped or oval. This can cause intermittent rubbing of the brake linings against the drum. See the preceding Section.
● Brake linings glazed. This condition, usually accompanied by heavy lining dust contamination, often induces brake squeal. The surface of the linings may be roughened using glass-paper or a fine file.

50 Brake induced fork judder

● Worn front fork stanchions and legs, or worn or badly adjusted steering head bearings. These conditions, combined with uneven or pulsating braking as described in Sections 44 and 48 will induce more or less judder when the brakes are applied, dependent on the degree of wear and poor brake operation. Attention should be given to both areas of malfunction. See the relevant Sections.

Electrical problems

51 Battery dead or weak

● Battery faulty. Battery life should not be expected to exceed 3 to 4 years. Gradual sulphation of the plates and sediment deposits will reduce the battery performance. Plate and insulator damage can often occur as a result of vibration. Complete power failure, or intermittent failure, may be due to a broken battery terminal. Lack of electrolyte will prevent the battery maintaining charge.
● Battery leads making poor contact. Remove the battery leads and clean them and the terminals, removing all traces of corrosion and tarnish. Reconnect the leads and apply a coating of petroleum jelly to the terminals.
● Load excessive. If additional items such as spot lamps, are fitted, which increase the total electrical load above the maximum alternator output, the battery will fail to maintain full charge. Reduce the electrical load to suit the electrical capacity.
● Rectifier failure.
● Alternator generating coils open-circuit or shorted.
● Charging circuit shorting or open circuit. This may be caused by frayed or broken wiring, dirty connectors or a faulty ignition switch. The system should be tested in a logical manner. See Section 54.

52 Battery overcharged

● Regulator faulty. Overcharging is indicated if the battery becomes hot or it is noticed that the electrolyte level falls repeatedly between checks. In extreme cases the battery will boil causing corrosive gases and electrolyte to be emitted through the vent pipes.
● Battery wrongly matched to the electrical circuit. Ensure that the specified battery is fitted to the machine.

53 Total electrical failure

● Fuse blown. Check the main fuse. If a fault has occurred, it must be rectified before a new fuse is fitted.

● Battery faulty. See Section 51.
● Earth failure. Check that the frame main earth strap from the battery is securely affixed to the frame and is making a good contact.
● Ignition switch or power circuit failure. Check for current flow through the battery positive lead (red) to the ignition switch. Check the ignition switch for continuity.

54 Circuit failure

● Cable failure. Refer to the machine's wiring diagram and check the circuit for continuity. Open circuits are a result of loose or corroded connections, either at terminals or in-line connectors, or because of broken wires. Occasionally, the core of a wire will break without there being any apparent damage to the outer plastic cover.
● Switch failure. All switches may be checked for continuity in each switch position, after referring to the switch position boxes incorporated in the wiring diagram for the machine. Switch failure may be a result of mechanical breakage, corrosion or water.
● Fuse blown. Refer to the wiring diagram to check whether or not a circuit fuse is fitted. Replace the fuse, if blown, only after the fault has been identified and rectified.

55 Bulbs blowing repeatedly

● Vibration failure. This is often an inherent fault related to the natural vibration characteristics of the engine and frame and is, thus, difficult to resolve. Modifications of the lamp mounting, to change the damping characteristics, may help.
● Intermittent earth. Repeated failure of one bulb, particularly where the bulb is fed directly from the generator, indicates that a poor earth exists somewhere in the circuit. Check that a good contact is available at each earthing point in the circuit.
● Reduced voltage. Where a quartz-halogen bulb is fitted the voltage to the bulb should be maintained or early failure of the bulb will occur. Do not overload the system with additional electrical equipment in excess of the system's power capacity and ensure that all circuit connections are maintained clean and tight.

HONDA MBX/MTX 125 & MTX 200

Check list

Daily (pre-ride checks)

1 Check the engine oil level
2 Check the coolant level
3 Ensure you have enough fuel to complete your journey
4 Check the battery electrolyte level
5 Check the operation of the brakes
6 Check the tyre pressures and tread condition
7 Check the suspension settings
8 Check, adjust and lubricate the final drive chain
9 Check that all controls operate smoothly and are well lubricated
10 Check for correct operation of the lights, horn, speedometer and turn signals

Four monthly, or every 2500 miles (4000 km)

1 Clean the air filter
2 Clean and re-gap the spark plug
3 Inspect the fuel and engine oil pipes for cracks or leakage and clean the fuel tap filters
4 Check the carburettor settings and the throttle and oil pump cable adjustment
5 Check the transmission oil level
6 Adjust the clutch cable
7 Check and adjust the brakes
8 Check the operation of the suspension and steering
9 Lubricate the control cables, stands and controls
10 Check the tightness of all fittings and fasteners

Eight monthly, or every 5000 miles (8000 km)

1 Decarbonise the engine and exhaust system
2 Inspect the cooling system components for signs of leakage or damage
3 Renew the spark plug
4 Check the wheels, bearings and tyres

Annually, or every 7500 miles (12 000 km)

1 Change the transmission oil
2 Clean the engine oil tank filter
3 Renew the brake fluid
4 Grease the steering head bearings
5 Renew the front fork oil
6 Dismantle, clean and grease the rear suspension
7 Grease the wheel bearings, speedometer drive and brake camshafts

Additional maintenance items – see text for intervals

1 Overhaul the hydraulic brake components
2 Renew the coolant
3 Clean the machine

Adjustment data

Spark plug gap
MBX125, MTX125 RW-F,
H, L, MTX200 RW-F   0.7 - 0.8 mm (0.028 - 0.032 in)
MTX125/200 RW-D   0.6 - 0.7 mm (0.024 - 0.028 in)

Idle speed
MBX/MTX125   1300 ± 100 rpm
MTX200   1400 ± 100 rpm

Final drive chain free play
MBX125   15 - 25 mm (0.6 - 1.0 in)
MTX125/200   35 - 45 mm (1.4 - 1.8 in)

Front fork air pressure   0 - 5.7 psi (0 - 0.4 kg/cm²)

Tyre pressures – front and rear
MBX125   32 psi (2.25 kg/cm²)
MTX125/200   21 psi (1.50 kg/cm²)*
Increase rear tyre pressure to 25 psi (1.75 kg/cm²) when carrying a pillion passenger

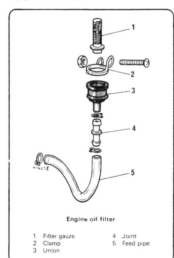

Engine oil filter

1 Filter gauze   4 Joint
2 Clamp   5 Feed pipe
3 Union

Recommended lubricants

	Component	Quantity	Type/viscosity
1	Engine	1.2 lit (2.1 pint)	Good quality 2-stroke engine oil suitable for injection system
2	Transmission	500 cc (0.9 pint)	SAE 10W/30 SE or SF engine oil
3	Cooling system 125 models 200 models	950 cc (1.67 pint) 1030 cc (1.81 pint)	30% corrosion-inhibited ethylene glycol-based anti-freeze. 70% distilled water
4	Air filter element	As required	SAE 80 or 90 gear oil
5	Front forks: MBX125 MTX125/200 RW-D MTX125 RW-F,H,L MTX200 RW-F	139.0 ± 2.5 cc (4.89 ± 0.09 fl oz) 311.5 ± 2.5 cc (10.97 ± 0.09 fl oz) 352 cc (12.39 fl oz) N/Av	Automatic transmission fluid (ATF)
6	Final drive chain: Standard chain O-ring chain	As required	Aerosol chain lubricant SAE 80 or 90 gear oil
7	Hydraulic front brake: MBX model MTX models	As required	SAE J1703 or DOT 3 brake fluid DOT 4 brake fluid
8	Wheel bearings	As required	High melting point grease
9	Speedometer drive	As required	High melting point grease
10	Steering head bearings	As required	General purpose grease
11	Hydraulic brake caliper axle bolts and pins	As required	PBC based caliper grease
12	Drum brake camshafts	As required	High melting point grease
13	Rear suspension pivots	As required	Molybdenum di-sulphide grease
14	Pivot points	As required	General purpose grease
15	Instrument drive cables	As required	General purpose grease
16	Control cables	As required	Light machine oil

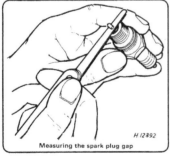

Measuring the spark plug gap

H.12392

ROUTINE MAINTENANCE GUIDE

Routine maintenance

Periodic routine maintenance is a continuous process which should commence immediately the machine is used. The object is to maintain all adjustments and to diagnose and rectify minor defects before they develop into more extensive, and often more expensive, problems.

It follows that if the machine is maintained properly, it will both run and perform with optimum efficiency, and be less prone to unexpected breakdowns. Regular inspection of the machine will show up any parts which are wearing, and with a little experience, it is possible to obtain the maximum life from any one component, renewing it when it becomes so worn that it is liable to fail.

Regular cleaning can be considered as important as mechanical maintenance. This will ensure that all the cycle parts are inspected regularly and are kept free from accumulations of road dirt and grime.

All tasks are grouped under various mileage headings, all of which are also given calendar-based intervals; if the machine only covers a low mileage maintenance should be carried out according to the calendar headings instead. All intervals are intended as a guide only; as a machine gets older it develops individual faults which require more frequent attention and if used under particularly arduous conditions it is advisable to reduce the period between each check.

For ease of reference, most service operations are described in detail under the relevant heading. However, if further general information is required, this can be found under the pertinent Section heading and Chapter in the main text.

Although no special tools are required for routine maintenance, a good selection of general workshop tools is essential. Included in the tools must be a range of metric ring or combination spanners, a selection of crosshead screwdrivers, and two pairs of circlip pliers, one external opening and the other internal opening. One further item of equipment that can be regarded as essential for owners of MTX125/ 200 models is a stand of some sort that can hold the machine securely in an upright position with enough height to permit the wheels to be removed. This can be anything from an old milk crate to a purpose-built paddock-type metal stand.

Daily (pre-ride check)

It is recommended that the following items are checked whenever the machine is about to be used. This is important to prevent the risk of unexpected failure of any component while riding the machine and, with experience, can be reduced to a simple checklist which will only take a few moments to complete. For those owners who are not inclined to check all items with such frequency, it is suggested that the best course is to carry out the checks in the form of a service which can be undertaken each week or before any long journey. It is essential that all items are checked and serviced with reasonable frequency.

1 Check the engine oil level

Although a warning lamp is fitted to give prior warning that the oil level is running low, it is only activated when there is 200cc or less oil remaining in the tank; this may not allow sufficient reserve to allow you to find some more. Accordingly it is recommended that the tank be kept topped up at all times; the true level can be seen through the translucent plastic of the tank. Remember that unlike a four-stroke, the lubrication system is total loss and the oil will be consumed at a steady rate depending on usage; if the tank is not refilled and the warning lamp malfunctions or is ignored, the first warning will be when the engine seizes up through lack of lubrication.

On MBX125 models unlock and remove the seat to gain access to the filler cap. On MTX125/200 models remove the left-hand side panel and swing out the tank extension. Unscrew the cap and add oil up to the bottom of the filler neck. Tighten the cap securely on refitting.

Use only a good quality two-stroke engine oil suitable for motorcycle injection systems; never use ordinary engine oil except in cases of real emergency and note that Honda specifically forbid the use of outboard engine (marine) two-stroke oils. Store a bottle of the correct oil with the machine as a reserve supply.

Engine oil level can be seen through the side of the tank – MTX125/200 shown

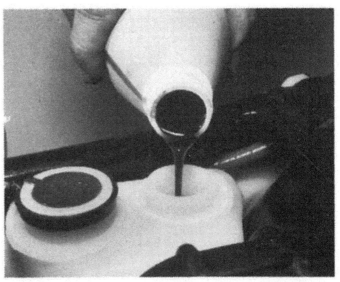

Use only good quality two-stroke oil when topping up – MBX125 shown

If the oil tank is ever allowed to run dry do not start the engine until it has been refilled and the oil feed pipes have been primed with the oil and bled to remove any air bubbles. If engine damage is suspected, it must be dismantled for checking and repair.

2 Check the coolant level

Although the cooling system is semi-sealed and should not require frequent topping up, it is still necessary to check the level at regular intervals. A separate expansion tank is fitted to allow for expansion of the coolant when the engine is hot, the displaced liquid being drawn back into the system when it cools. It is therefore the level of coolant in the expansion tank which is to be checked; the tank is constructed of translucent plastic so that the coolant level can be seen easily in relation to the upper and lower level lines marked on the tank side.

On MBX125 models the tank is on the right-hand side immediately above and to the rear of the swinging arm pivot; it will be necessary to remove the side panel to see the level marks clearly and/or to top up. On MTX125/200 models the tank is inside the left-hand radiator shroud, adjacent to the fuel tank.

The coolant level must be between the higher ('Full' or 'F') and lower ('Low' or 'L') level marks at all times. Although the level will vary with engine temperature, Honda state that if the level is ever found to be below the lower level mark it must be topped up to the higher level mark regardless of engine temperature. If the level is significantly above the higher level mark at any time the surplus should be siphoned off to prevent coolant being blown over the rear of the machine via the tank breather.

Use only the specified ingredients to make coolant of the required strength, as described in Chapter 2, and always have a supply prepared for topping up. In cases of real emergency distilled water or **clean** rainwater may be used, but remember that this will dilute the coolant and reduce the degree of protection against freezing.

If the level falls steadily, check the system very carefully for leaks, as described in Chapter 2. Also, do not forget to check that the radiator matrix is clean, unblocked and free from damage of any sort; this applies particularly to owners of MTX125/200 models who have been riding off-road.

3 Check the fuel level

Checking the petrol level may seem obvious, but it is all too easy to forget. Ensure that you have enough petrol to complete your journey, or at least to get you to the nearest petrol station.

4 Check the battery

The battery is located on the right-hand side of the machine, to the rear of the engine. On MBX125 models remove the side panel to check the electrolyte level; if the battery is to be removed disconnect the terminals (negative terminal first, always) and remove the nut retaining both parts of the retaining strap. On MTX125/200 models the electrolyte level can be seen without the need for preliminary dismantling; if the battery is to be removed unscrew the retaining bolt so that the cover can be withdrawn, the battery partially removed and the terminals disconnected.

On all models, whenever the battery is disconnected, remember to disconnect the negative (–) terminal first, to prevent the possibility of short circuits. The electrolyte level, visible through the translucent casing, must be between the two level marks. If necessary remove the cell caps and top up to the upper level using only distilled water. Check that the terminals are clean and apply a thin smear of petroleum jelly (not grease) to each to prevent corrosion. On refitting, check that the vent hose is not blocked and that it is correctly routed with no kinks, also that it hangs well below any other component, particularly the chain or exhaust system. Remember always to connect the negative (–) terminal last when refitting the battery.

Always check that the terminals are tight and that the rubber covers are correctly refitted, also that the fuse connections are clean and tight, that the fuse is of the correct rating and in good condition, and that a spare is available on the machine should the need arise.

At regular intervals remove the battery and check that there is no pale grey sediment deposited at the bottom of the casing. This is caused by sulphation of the plates as a result of re-charging at too high a rate or as a result of the battery being left discharged for long periods. A good battery should have little or no sediment visible and its plates should be straight and pale grey or brown in colour. If sediment deposits are deep enough to reach the bottom of the plates, or if the plates are buckled and have whitish deposits on them, the battery is faulty and must be

renewed. Remember that a poor battery will give rise to a large number of minor electrical faults.

If the machine is not in regular use, disconnect the battery and give it a refresher charge every month to six weeks, as described in Chapter 7.

5 Check the brakes

Check that the front and rear brakes work effectively and without binding. Ensure that the rod linkages and the cables, as applicable, are lubricated and properly adjusted. Check the fluid level in the master cylinder reservoir, where appropriate, and ensure that there are no fluid leaks. Should topping-up be required, use only the recommended hydraulic fluid. Note that the brake lever on hydraulically-operated brakes should have free play of 10 – 20 mm (0.4 – 0.8 in) measured at the lever ball end. If there is any more or less than this amount, the brake system should be checked as described in Chapter 6 or Routine Maintenance.

The height of the rear brake pedal can be adjusted by means of a stop bolt and locknut fitted at the pedal rear end. Although the final setting is up to the individual owner's requirement, it is recommended the pedal tip is 10 – 20 mm (0.4 – 0.8 in) below the level of the footrest top surface. When adjustment is correct, tighten securely the locknut; if the pedal height was altered significantly check the rear brake and stop lamp rear switch settings.

Coolant level is checked at expansion tank – MTX125/200 shown

If level is low, top up expansion tank ...

... to higher level mark – MBX125 shown

Electrolyte level must be between level marks on casing

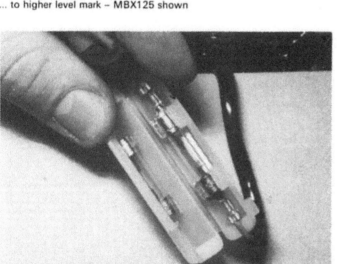

Check that fuse and connections are in good condition – always have a spare fuse ready

Front brake hydraulic fluid level must be kept above lower level mark

Brake pedal height adjustment is made at stop bolt shown

6 Check the tyre pressures and tread wear

Check the tyre pressures with a gauge that is known to be accurate. It is worthwhile purchasing a pocket gauge for this purpose because the gauges on garage forecourt airlines are notoriously inaccurate. The pressures, which should be checked with the tyres cold, are 32 psi (2.25 kg/cm²) for MBX125 models and 21 psi (1.50 kg/cm²) for MTX125/200 models. The pressures are the same for front and rear tyres and for solo or pillion use except for MTX125/200 models, where the rear tyre pressure should be increased to 25 psi (1.75 kg/cm²) when a pillion passenger is carried.

At the same time as the tyre pressures are checked, examine the tyres themselves. Check them for damage, especially splitting of the sidewalls. Remove any small stones or other road debris caught between the treads. When checking the tyres for damage, they should be examined for tread depth in view of both the legal and safety aspects. It is vital to keep the tread depth within the UK legal limits of 1 mm of depth over three-quarters of the tread breadth around the entire circumference with no bald patches. Many riders, however, consider nearer 2 mm to be the limit for secure roadholding, traction, and braking, especially in adverse weather conditions, and it should be noted that Honda recommend minimum tread depths of 1.5 mm (0.06 in) for the front tyre and 2.0 mm

(0.08 in) for the rear on MBX125 models, 3.0 mm (0.12 in) front and rear on MTX125/200 models; these measurements to be taken at the centre of the tread. Renew any tyre that is found to be damaged or excessively worn.

7   Check the suspension settings

The front forks are provided with an air valve in the top of each leg to enable the fork's effective spring rate to be altered to suit the individual owner's needs. The pressure must be checked frequently to ensure that it is at the desired setting and that it is exactly the same for both legs.

Two tools are essential for setting for air pressure; a gauge capable of reading the low pressures involved, and a low-pressure pump. The gauge must be finely calibrated to ensure that both legs can be set to the same pressure, and must cause only a minimal drop in pressure whenever a reading is taken; as the total air volume is so small, an ordinary gauge, such as a tyre pressure gauge, will cause a large drop in pressure because of the amount of air required to operate it. Gauges for use on suspension components are now supplied by several companies and should be available through any good motorcycle dealer. The pump must be of the hand- or foot-operated type, a bicycle pump being ideal; aftermarket pumps are available for use on suspension systems and are very useful, but expensive. **Never** use a compressor-powered air line; it is all too easy to exceed the maximum recommended pressure which may cause damage to the fork oil seals and may even result in personal injury. Add air very carefully, by means of a hand-operated pump only, and in small amounts at a time.

The air pressure must be set when the forks are cold, and with the machine securely supported so that its front wheel is clear of the ground, thus ensuring that the air pressure is not artificially increased.

Set the pressure to the required amount within the specified range of 0 – 5.7 psi (0 – 0.4 kg/cm²) then be careful to ensure that each leg is at exactly the same pressure within the tolerance of 1.5 psi (0.1 kg/cm²). This is essential as any imbalance in pressures will impair fork performance and may render the machine unsafe to ride. Note that good quality aftermarket kits are now available to link separate air caps; the use of one of these, when correctly installed, will ensure that the pressures in the legs are equal at all times and will aid the task of setting the pressure in the future.

It is permissible to experiment with air pressures until the ideal setting is found, but this will require a great deal of care and patience. Do not exceed the maximum recommended pressure of 42.7 psi (3.0 kg/cm²).

Other than the very coarse method of spring preload adjustment on MTX125/200 models the rear suspension is not variable.

8   Check, adjust and lubricate the final drive chain

The exact interval at which the final drive chain will require lubrication adjustment and renewal is entirely dependent on the usage to which the machine is put and on the amount of care devoted to chain maintenance. In some cases the chain will require daily lubrication, in other cases it need only be lubricated at weekly intervals. The best rule to follow is that if the chain rollers look dry, then the chain needs lubrication immediately. Do not allow the chain to run dry until the links start to kink, or until traces of reddish-brown deposit can be seen on the sideplates. For ease of reference the full procedure is given here but it must be up to the owner to decide how often the chain needs attention.

Note that two types of chain are now available. Standard chains are packed with grease on assembly at the factory and, in addition to normal lubrication, must be removed from the machine at regular intervals so that they can be cleaned thoroughly and re-packed with chain grease as described below. O-ring chains are wider and have small O-rings set between the inner and outer sideplates of each link to seal the grease into the bearings for the life of the chain. These are easily identified by the presence of the O-rings and require a special approach in some aspects of maintenance. Although fitted as standard to some MTX200 models only, O-ring chains are increasingly popular due to their greatly-increased life and may be encountered on other models.

Cleaning

Although dirt and old lubricant may be washed off the chain while it is in place on the machine, this is not really satisfactory, especially with standard chains where the grease may be washed out of the bearings. To clean the chain thoroughly disconnect it at its connecting link and remove it from the machine. Note that refitting the chain is greatly simplified if a worn out length is temporarily connected to it. As the original chain is pulled off the sprockets, the worn-out chain will follow

it and remain in place while the task of cleaning and examination is carried out. On reassembly, the process is repeated, pulling the worn-out chain over the sprockets so that the new chain, or the freshly cleaned and lubricated chain, is pulled easily into place.

To clean the chain, immerse it in a bath containing a mixture of petrol and paraffin and use a stiff-bristled brush to scrub away all the traces of dirt and old lubricant. Take the necessary fire precautions when using this flammable solvent. Swill the chain around to ensure that the solvent penetrates fully into the bushes and rollers and can remove any lubricant which may still be present. When the chain is completely clean, remove it from the bath and hang it up to dry.

Refitting a new, or freshly-lubricated, chain is a potentially messy affair which is greatly simplified by connecting it to the worn-out length of chain used during its removal to pull it around the sprockets. Refit the connecting link, ensuring that the spring clip is fitted with its closed end facing the normal direction of travel of the chain.

Where an O-ring chain is fitted, take care not to lose the O-rings when the connecting link is dismantled, and clean it using kerosene (paraffin) only. Do not use any strong solvents as these may damage the O-rings and never use a steam cleaner or pressure washer. Lubricate the chain thoroughly with gear oil after cleaning and drying. On reassembly, ensure that the four O-rings are positioned correctly and ensure that there is no space between the connecting link sideplate and spring clip.

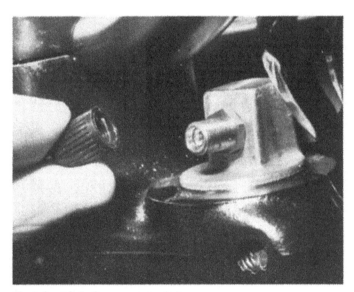

Check front fork air pressure regularly

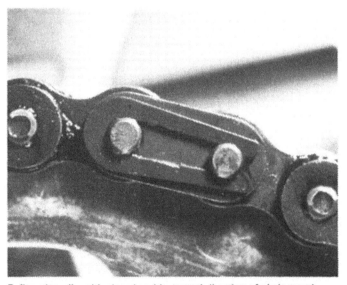

Refit spring clip with closed end in normal direction of chain travel

Checking for wear

The chain consists of a multitude of small bearing surfaces which will wear rapidly, and expensively, if the chain is not regularly lubricated and adjusted. A simple check for wear is as follows. With the chain fully lubricated and correctly adjusted as described below, attempt to pull the chain backwards off the rear sprocket. If the chain can be pulled clear of the sprocket teeth it must be considered worn out and renewed. This is the only practical test of wear for O-ring chains although in some cases the manufacturer provides red warning labels on the adjusters to indicate the need for chain renewal when, with the chain correctly adjusted, the red warning area aligns with the fork end or adjuster stop pin.

A more accurate measurement of chain wear involves the removal of the chain from the machine as described above. It applies principally to standard-type chains as the presence of the grease in an O-ring chain may take up free play which cannot, therefore, be measured.

To access accurately the amount of wear present in the chain, it must be cleaned and dried as described above, then laid out on a flat surface. Compress the chain fully and measure its length from end to end. Anchor one end of the chain and pull on the other end, drawing the chain out to its fullest extent. Measure the stretched length. If the stretched measurement exceeds the compressed measurement by more than 2%, or $^1/_4$ in per foot, the chain must be considered worn out and be renewed. Honda's own recommendation is that the maximum length of any 40 links (ie mark any one pin, count off 41 pins and measure the distance between the two) must not exceed the service limits given in the Specifications Section of Chapter 1 when the chain is cleaned and stretched out.

Renewal

If the chain is found to be worn out, or if any links are kinked or stiff through lack of lubrication, or if damage such as split or missing rollers or side plates is found, the chain must be renewed. This should be done always in conjunction with both sprockets since the running together of new and part-worn components greatly increases the rate of wear of all three items, resulting in even greater expenditure than necessary.

Note that replacement chains are available in standard metric sizes from Renold Ltd, the British chain manufacturer. When purchasing a new chain always quote the size, the number of links required and the machine to which the chain is to be fitted. **Standard** chain sizes are given in the Specifications Section of Chapter 1, but note that the gearing may well have been altered on some machines.

Lubrication

On standard type chains, for the purpose of daily or weekly lubrication, one of the many proprietary aerosol-applied chain lubricants can be applied, while the chain is in place on the machine. It should be applied at least once a week, and daily if the machine is used in wet weather conditions. Engine oil can be used for this task, but remember that it is flung off the chain far more easily than grease, thus making the rear end of the machine unnecessarily dirty, and requires more frequent application if it is to perform its task adequately. Also remember that surplus oil will eventually find its way on to the tyre, with quite disastrous consequences. While this will serve as a stop-gap measure, it does nor reach the inner bearing surfaces of the chain. These can be lubricated correctly only by removing and cleaning the chain as described above and by immersing it in a molten bath of special chain lubricant such as Chainguard or Linklyfe; this should be done at intervals of 500 – 1000 miles for normal road use and more frequently if the machine is used in wet weather or poor conditions. Follow carefully the manufacturer's instructions when using Chainguard or Linklyfe and take great care to swill the chain gently in the molten lubricant to ensure that it penetrates to all bearing surfaces. Wipe off the surplus before refitting the chain to the machine.

Although grease is sealed into the inner bearings of O-ring chains, removing the need for regular immersion in molten lubricant, these chains must still be lubricated at daily or weekly intervals to keep the O-rings supple and to prevent wear between the rollers and sprocket teeth. With the machine supported so that the rear wheel is clear of the ground, revolve the wheel and allow oil to dribble on to the rollers and between the sideplates until the chain is completely oily all along its length. Use only SAE80 or 90 gear oil for this purpose. **Warning:** most aerosol lubricants contain chemicals which will attack the O-rings causing loss of sealed-in lubricant and premature chain wear. Check carefully that an aerosol lubricant is marked as being suitable for use with O-ring chains before applying it; some have appeared on the market in recent years.

Adjustment

It is necessary to check the chain tension at regular intervals to compensate for wear. Since this wear does not take place evenly along the length of the chain, tight spots will appear which must be compensated for when adjustment is made. Chain tension is checked with the transmission in neutral and with the machine standing on its wheels. Find the tightest spot on the chain by pushing the machine along and feeling the amount of free play present on the bottom run of the chain, midway between the sprockets, testing along the entire length of the chain. Press the chain tensioner, where fitted, away from the chain to ensure that free play is measured accurately. When the tightest spot has been found, measure the total amount of up and down movement available; this should be between 15 – 25 mm (0.6 – 1.0 in) on MBX125 models, 35 – 45 mm (1.4 – 1.8 in) on MTX125/200 models. Note that chain free play must never be allowed to exceed 50 mm (2.0 in) on MBX125 models, 60 mm (2.4 in) on MTX125/200 models; this could cause serious damage by allowing the chain to rub against the frame.

If adjustment is necessary on MBX125 and MTX200 RW-F models, slacken the rear wheel spindle nut by just enough to permit the spindle to be moved, slacken the adjuster locknuts and tighten by an equal amount both adjuster nuts to draw the spindle back the necessary amount. To preserve accurate wheel alignment, ensure that the same notches stamped in the adjusters line up with the same lines stamped on each swinging arm fork end. On MTX125/200 RW-D and MTX125 RW-F, H, L models slacken the rear wheel spindle nut by just enough to permit the spindle to be moved, then draw the spindle back by rotating the snail cams to the same amount on both sides. Use the numbers stamped in the cams to ensure that the same cam cutout engages with the stopper pin in each swinging arm fork end, thus preserving accurate wheel alignment.

On all models, a final check of accurate wheel alignment can be made by laying a plank of wood or drawing a length of string parallel to the machine so that it touches both walls of the rear tyre. Wheel alignment is correct when the plank or string is equidistant from both walls of the front tyre when tested on both sides of the machine, as shown in the accompanying illustration.

When the chain is correctly tensioned, apply the rear brake to centralise the shoes on the drum and tighten the spindle retaining nut securely. Use the recommended torque settings where possible and do not forget to tighten the adjuster locknuts, where appropriate. Note that if the chain tension has been altered significantly, the rear brake and stop lamp rear switch adjustment will also require resetting; these should be checked before taking the machine out on the road.

Finally, on machines so equipped, check that the chain tensioner rubbing block is in good condition, renewing it if worn excessively, and that the tensioner arm moves freely against spring pressure. Apply a few drops of oil to the tensioner pivot and renew the spring if it appears to have weakened or if it is damaged. Check also that the chain guides and swinging arm pivot guide (where fitted) are straight and unworn. Renew them if they are badly worn by the action of the chain.

Check chain tension in position shown – find tightest spot in chain

Slacken locknut and tighten adjuster nut to tighten chain ...

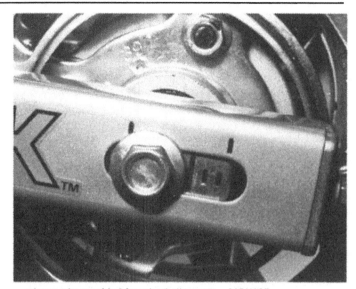

... using marks provided for wheel alignment – MBX125, MTX200 RW-F

Other MTX125/200 models – use numbered cutouts in snail cams to ensure wheel is aligned

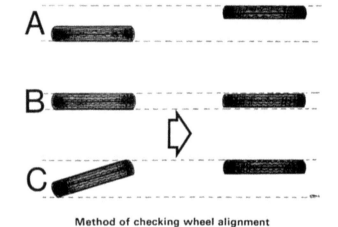

Method of checking wheel alignment

A & C –   Incorrect   B – Correct

9   Check the controls

Check the throttle and clutch cables and levers, the gear lever and the footrests to ensure that they are adjusted correctly, functioning correctly, and that they are securely fastened. If a bolt is going to work loose, or a cable snap, it is better that it is discovered at this stage with the machine at a standstill, rather than when it is being ridden.

10 Legal check

Check that all lights, turn signals, horn and speedometer are working correctly to make sure that the machine complies with all legal requirements in this respect. Check also that the headlamp is correctly aimed. Rotate the spring-loaded adjusting screw to alter the horizontal alignment; the vertical aim must be aligned so that with the machine standing on its wheels on level ground with the rider (and pillion passenger, if one is regularly carried) seated normally, the dip beam centre (as shown on a wall 25 feet away) must be at the same height from the ground as the centre of the headlamp itself. This is adjusted, after removing the headlamp nacelle or fairing by slackening the headlamp mounting bolts and tilting the headlamp to the correct angle. Note that reference marks are stamped on the headlamp and mounting bracket to provide an initial setting.

Rotate adjusting screw for horizontal aim – slacken mounting bolt and tilt headlamp for vertical aim

Four monthly, or every 2500 miles (4000 km)

Carry out all checks given under the daily heading, particularly those relating to the battery, coolant level and drive chain, then carry out the following:

1 Clean the air filter

Remove the left-hand side panel and unscrew the air filter cover retaining screws, then withdraw the filter cover. On MBX125 models pull out the spring retaining clip, remove the element assembly and peel the element off its supporting frame. On MTX125/200 models pull the filter assembly out of its locating slots and separate the two supporting frames from the element.

Check that the element is not split, hardened or deteriorated through age or severe clogging; renew it if it is damaged in this way. Soak it in a non-flammable or high flash-point solvent such as white spirit (petrol is not recommended because of the fire risk) squeezing it gently to remove all old oil and dirt. Remove excess solvent by squeezing the element between the palms of the hands; wringing it out will damage it. Put the element to one side to allow any remaining solvent to evaporate.

When it is completely dry, soak the element in clean SAE 80 or 90 gear oil and gently squeeze out any surplus to leave it only slightly oily to the touch. Refit the element to its frame.

On reassembly, ensure that the element is correctly seated and is secured by its retaining clip (MBX125). On MTX125/200 models hold the assembly together to form a sandwich and fit it into its housing slots with the narrower edge inwards. Refit the cover, checking that the sealing O-ring (where fitted) is in good condition and seated correctly and applying a smear of grease to the cover sealing edges to provide additional protection, then tighten the retaining screws securely. Refit the side panel.

Remember that the filter element and cover must be positioned correctly to provide a good seal so that unfiltered air cannot carry dirt into the engine, also that the carburettor is jetted to compensate for the pressure of the air filter; serious damage will occur to the engine through overheating as a result of the weak mixture which will result if the filter is damaged or bypassed in this way. For these reasons the engine should never be run with the air filter removed or disconnected.

This interval is the maximum for filter cleaning; if the machine is used in wet weather or in very dirty or dusty conditions, the filter must be cleaned much more frequently.

2 Check the spark plug

The spark plug supplied as original equipment will prove satisfactory in most operating conditions; alternatives are available to allow for varying altitudes, climatic conditions and the use to which the machine is put. If the spark plug is suspected of being faulty it can be tested only by the substitution of a brand new (not second-hand) plug of the correct make, type, and heat range; always carry a spare on the machine.

Air filter cleaning MBX125 – remove filter cover ...

... withdraw spring clip ...

... and pull out filter element assembly

Ensure spark plug electrodes are cleaned and gapped correctly

Note that the advice of a competent Honda Service Agent or similar expert should be sought before the plug heat range is altered from standard. The use of too cold, or hard, a grade of plug will result in fouling and the use of too hot, or soft, a grade of plug will result in engine damage due to excess heat being generated. If the correct grade of plug is fitted, however, it will be possible to use the condition of the spark plug electrodes to diagnose a fault in the engine or to decide whether the engine is operating efficiently or not. The accompanying series of colour photographs will show this clearly. Also, always ensure that the plug is of the resistor type (indicated by the letter 'R') so that its resistance value is correct for the ignition system. The same applies to the suppressor cap; if a cap or plug of the wrong type is fitted, thus producing a much greater or lesser resistance value than that for which the ignition system was designed, one or more components of the system may break down.

It is advisable to carry a new spare spark plug on the machine, having first set the electrodes to the correct gap. Whilst spark plugs do not often fail, a new replacement is well worth having if a breakdown does occur. Ensure that the spare is of the correct heat range and type.

The electrode gap can be assessed using feeler gauges. If necessary, alter the gap by bending the outer electrode, preferably using a proper electrode tool. **Never** bend the centre electrode, otherwise the ceramic insulator will crack, and may cause damage to the engine if particles break away whilst the engine is running. If the outer electrode is seriously eroded as shown in the photographs, or if the spark plug is heavily fouled, it should be renewed. Clean the electrodes using a wire brush or a sharp-pointed knife, followed by rubbing a strip of fine emery across the electrodes. If a sand-blaster is used, check carefully that there are no particles of sand trapped inside the plug body to fall into the engine at a later date. For this reason such cleaning methods are no longer recommended; if the plug is so heavily fouled it should be renewed.

Before refitting a spark plug into the cylinder head; coat the threads sparingly with a graphited grease to aid future removal. Use the correct size spanner when tightening the plug, otherwise the spanner may slip and damage the ceramic insulator. The plug should be tightened by hand only at first and then secured with a quarter turn of the spanner so that it seats firmly on its sealing ring.

Never overtighten a spark plug otherwise there is risk of stripping the thread from the cylinder head, especially as it is cast in light alloy. A stripped thread can be repaired without having to scrap the cylinder head by using a 'Helicoil' thread insert. This is a low-cost service, operated by a number of dealers.

3 Check the fuel and engine oil lines and clean the fuel tap filters

Switch the tap to the 'Off' position and unscrew the filter bowl from the tap base, then remove the O-ring and filter gauze. Check the condition of the sealing O-ring and renew, it if it is seriously compressed, distorted or damaged. Clean the filter gauze using a fine-bristled toothbrush or similar; remove all traces of dirt or debris and renew the gauze if it is split or damaged. Thoroughly clean the filter bowl; if excessive signs of dirt or water are found in the petrol, remove the tank as described in Chapter 3, empty the petrol into a clean container and remove the fuel tap by unscrewing its retaining gland nut.

Remove the tubular filter gauze from the tap stack pipe, noting the presence of a small spacer and of the sealing O-ring, and clean it using a fine-bristled brush; if the gauze is split, twisted or damaged it should be renewed. Flush the tank until all traces of dirt or water are removed. The tap cannot be dismantled further and must be renewed as a complete assembly if the lever is leaking or defective in any way. If its passages are blocked, use compressed air to blow them clear.

On reassembly, renew the sealing O-ring if damaged or worn. Fit the filter gauze, spacer and O-ring into the tap and refit the assembly to the tank. Check that the tap is correctly aligned before tightening the gland nut; do not overtighten the gland nut or its threads may be stripped, necessitating the renewal of the tap.

Fit the filter gauze to the tap, ensuring that it is located correctly, then press the O-ring into place to retain it. Use only a close-fitting ring spanner to tighten the filter bowl, which should be secured by just enough to nip up the O-ring; do not overtighten it as this will only damage the filter bowl, distort the O-ring and promote fuel leaks. The recommended torque setting is only 0.3 - 0.5 kgf m (2 - 3.5 lbf ft). If any leaks are found in the tap they can be cured only by the renewal of the tap assembly or the defective seal, where separate.

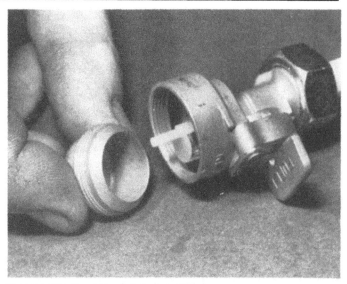
Unscrew filter bowl to reach fuel tap filter gauze

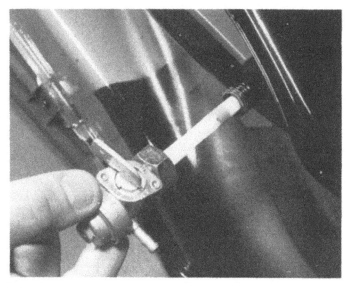
It may be necessary to remove tap so that upper filter can be cleaned

Ensure filter gauze is located correctly in tap body ...

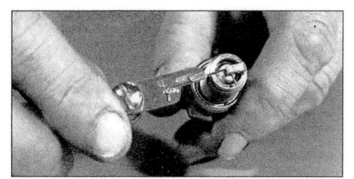

Electrode gap check - use a wire type gauge for best results

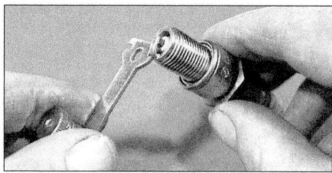

Electrode gap adjustment - bend the side electrode using the correct tool

Normal condition - A brown, tan or grey firing end indicates that the engine is in good condition and that the plug type is correct

Ash deposits - Light brown deposits encrusted on the electrodes and insulator, leading to misfire and hesitation. Caused by excessive amounts of oil in the combustion chamber or poor quality fuel/oil

Carbon fouling - Dry, black sooty deposits leading to misfire and weak spark. Caused by an over-rich fuel/air mixture, faulty choke operation or blocked air filter

Oil fouling - Wet oily deposits leading to misfire and weak spark. Caused by oil leakage past piston rings or valve guides (4-stroke engine), or excess lubricant (2-stroke engine)

... and refit seal, as shown to retain it

Use only close-fitting spanner to prevent filter bowl damage – do not overtighten

Give the pipe which connects the fuel tap and carburettor a close visual examination and check for cracks or any signs of leakage. In time, the synthetic rubber pipe will tend to deteriorate, and will eventually leak. Apart from the obvious fire risk, the leaking fuel will affect fuel economy. The pipe will usually split only at the ends; if there is sufficient spare length the damaged portion can be cut off and the pipe refitted. The seal is effected by the interference fit of the pipe on the spigot; although the wire clips are only an additional security measure they should always be refitted correctly and should be renewed if damaged, twisted or no longer effective. If the pipe is to be renewed, always use the correct replacement type and size of neoprene tubing to ensure a good leak-proof fit. Never use natural rubber tubing, as this breaks up when in contact with petrol and will obstruct the carburettor jets, or clear plastic tubing which stiffens to the point of being brittle when in contact with petrol and will produce leaks that are difficult to cure.

Working as described for the fuel lines, check the engine oil feed lines very carefully. Check that the fitting clamps and wire clips are securely fastened and that the pipes are secured by any clamps or ties provided; ensure that no part of the tubing is chafing on any part of the engine or frame. Any damaged piping must immediately be renewed. If signs of oil leakage or if air bubbles or dirt in the pipes are found, the fault must be cured immediately.

4 Check the carburettor settings and throttle and oil pump cable adjustment

If rough running of the engine has developed, some adjustment of the carburettor pilot setting and tick over speed may be required. If this is the case refer to Chapter 3 for details. Do not make these adjustments unless they are obviously required; there is little to be gained by unwarranted attention to the carburettor. Complete carburettor maintenance by removing the drain plug on the float chamber, turning the petrol on, and allowing a small amount of fuel to drain through, thus flushing any water or dirt from the carburettor. Refit the drain plug securely and switch the petrol off.

Once the carburettor has been checked and reset if necessary, the throttle cable free play can be checked. Open and close the throttle several times, allowing it to snap shut under its own pressure. Ensure that it is able to shut off quickly and fully at all handlebar positions, then check that there is 2 - 6 mm (0.08 - 0.24 in) free play measured in terms of twistgrip rotation (ie 10 - 15°). Use the adjuster at the twistgrip to achieve the correct setting then open and close the throttle again to settle the cable and check that the adjustment has not altered.

Removing its three retaining screws, withdraw the plastic oil pump cover from the front of the crankcase right-hand cover and note the

alignment mark on the pump control lever. Open the throttle fully and check that the reference mark on the pump control aligns exactly with the fixed index mark stamped on a projection on the pump body. If adjustment is necessary, slacken the adjuster locknut and turn the adjusting nut until the marks line up exactly; remember that the throttle must be fully open to align the marks. Once adjustment is correct, open and close the throttle several times to settle the cable and to ensure that the lever moves smoothly and steadily throughout its full travel. Check that the adjustment has remained the same, apply a few drops of oil to all exposed lengths of inner cable and to the lever pivots, then refit the oil pump cover.

If the oil pump is to be correctly synchronised with the carburettor to ensure that the engine receives the correct oil supply, the oil pump cable adjustment must be checked, and reset if necessary, after any work of any type is carried out on the carburettor or throttle cable.

Remove oil pump cover to check cable adjustment

Pump control lever reference mark must align exactly with fixed index mark at full throttle

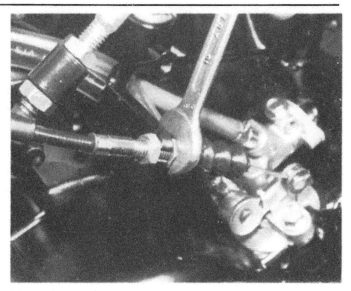

Adjust oil pump cable at adjuster shown

5 Check the transmission oil level

Start the engine and allow it to warm up to normal operating temperature so that the oil is at the true level. Stop the engine, wait 2-3 minutes for the level to settle, then unscrew the level plug from the crankcase right-hand cover. This is of a different type from the cover fastening screws and is situated below the kickstart lever and to the rear of the clutch bulge.

With the machine standing absolutely upright on its wheels on level ground, the transmission oil level is correct if oil just trickles out of the level plug orifice. If the flow is too fast, check that the machine is upright, then allow the surplus to drain out until it slows to a trickle; if little or no oil appears, remove the filler plug from the front of the cover and add oil until the level is correct. Use only a good quality SAE 10W30 SE or SF engine oil.

Renew its sealing washer if damaged or worn and refit the level plug (and filler plug, if removed), tightening it securely.

6 Adjust the clutch

The clutch is adjusted correctly when there is 10 – 20 mm (0.4 – 0.8 in) of free play in the cable, measured at the handlebar lever ball end, and the clutch operates smoothly with no sign of slip or drag.

There is no provision for adjustment of the clutch itself or of the release mechanism; adjustment can only be made at the cable. Normal adjustment is made only at the cable lower end, reserving the handlebar adjuster for quick roadside adjustments. Slacken the adjuster locknuts and screw in fully the handlebar adjuster (if necessary), then turn the lower adjusting nut until the correct setting is achieved. Tighten both adjuster locknuts and apply a few drops of oil to all exposed lengths of inner cable and all lever pivots and cable nipples.

If adjustment is used up on the lower adjuster, both adjusters may be used together; if both are used up the cable must be renewed. If cable adjustment does not eliminate slip or drag, or if any other problems are encountered, the clutch must be dismantled for examination as described in Chapter 1.

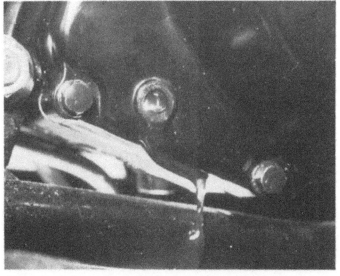

Oil should trickle slowly out of level plug orifice if level is correct

Add oil via filler plug orifice shown

Adjusting the clutch cable

Do not allow fluid level above upper level line (arrowed)

Diaphragm should be cleaned, dried and folded as shown before refitting

7 Check and adjust the brakes

The hydraulic front brake fitted to MBX125, MTX125 RW-F, H, L and MTX200 RW-F models requires no regular adjustments; pad wear is compensated for by the automatic entry of more fluid into the system from the handlebar reservoir. All that is necessary is to maintain a regular check on the fluid level and the degree of pad wear.

To check the fluid level, turn the handlebars until the reservoir is horizontal and check that the fluid level, as seen through the sight glass in the rear face of the reservoir body, is not below the lower level mark on the body. Remember that while the fluid level will fall steadily as the pad friction material is used up, if the level falls below the lower level mark there is a risk of air entering the system; it is therefore sufficient to maintain the fluid level above the lower level mark, by topping-up if necessary. Do not top up to the higher level mark (formed by a cast line on the inside of the reservoir) unless this is necessary after new pads have been fitted. If topping up is necessary, wipe any dirt off the reservoir, remove the retaining screws and lift away the reservoir cover and diaphragm. Use only good quality brake fluid of the recommended type and ensure that it comes from a freshly opened sealed container; brake fluid is hygroscopic, which means that it absorbs moisture from the air, therefore old fluid may have become contaminated to such an extent that its boiling point has been lowered to an unsafe level. Remember also that brake fluid is an excellent paint stripper and will attack plastic components; wash away any spilled fluid immediately with copious quantities of water. When the level is correct, clean and dry the diaphragm, fold it into its compressed state and fit it to the reservoir. Refit the reservoir cover (and gasket, where fitted) and tighten securely, but do not overtighten, the retaining screws.

To check the degree of pad wear, look closely at the pads from above or below the caliper. Wear limit marks are provided in the form of deep notches cut in the top and bottom edges of the friction material or as red painted lines cut around the outside of the material. If either pad is worn at any point so that the inside end of the mark (next to the metal backing) is in contact with the disc, or if the wear limit marks have been removed completely, both pads must be renewed as a set. If the pads are so fouled with dirt that the marks cannot be seen, or if the oil or grease is seen on them, they must be removed for cleaning and examination.

To remove the pads, use an Allen key of suitable size to unscrew the threaded plugs which seal the pad retaining pins, then slacken the retaining pins. Remove the two mounting bolts and withdraw the caliper, complete with its mounting bracket, from the fork leg, taking care not to twist the brake hose. Remove both pad retaining pins and withdraw the pads, noting carefully the presence and exact positions of the anti-rattle spring (all models) and the retainer and anti-squeal shim (MTX125/200 models).

Do not overtighten cover retaining screws

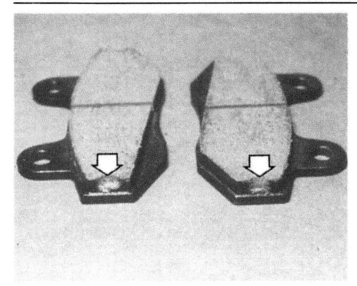

Brake pads must be renewed when wear limit marks (arrowed) are worn away

To remove brake pads, unscrew threaded plugs ...

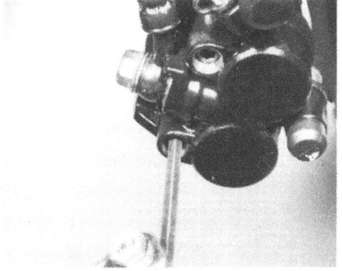

... and slacken pad retaining pins ...

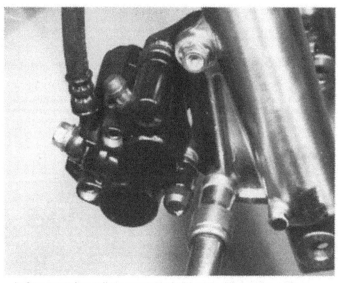

... before removing caliper mounting bolts and withdrawing caliper

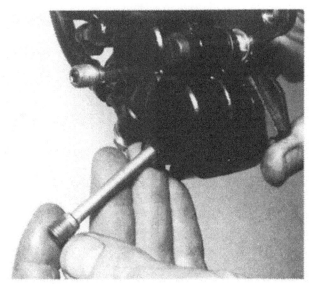

Remove pad retaining pins to release brake pads

Thoroughly clean all components, removing all traces of dirt, grease and old friction material then use carefully fine abrasive paper to polish clean any corroded items. Check carefully that the caliper body slides easily on its two axle bolts and that there is no damage to any of the caliper components, especially the rubber seals. If it is stiff, remove the mounting bracket from the caliper, clean the axle bolts and check them for wear or damage (which can be cured only by the renewal of the bolts or mounting bracket, as applicable) then smear a good quantity of silicone or PBC (Poly Butyl Cuprysil) based caliper grease over the bolts and caliper bores before refitting the mounting bracket to the caliper. It is essential that the caliper body can move smoothly and easily on the mounting bracket for the brake to be effective. **Warning:** do not use ordinary high-melting point grease; this will melt and foul the pads, rendering the brake ineffective.

If the pads are worn to the limit marks, fouled with oil or grease, or heavily scored or damaged by dirt and debris, they must be renewed as a set; there is no satisfactory way of degreasing friction material. If the pads can be used again, clean them carefully using a fine wire brush that is completely free of oil or grease. Remove all traces of road dirt and corrosion, then use a pointed instrument to clean out the groove(s) in the friction material and to dig out any embedded particles of foreign matter. Any areas of glazing may be removed using emery cloth.

Caliper must be able to slide easily on mounting bracket axle pins –
clean and grease

On reassembly, if new pads are to be fitted, the caliper pistons must
now be pushed back as far as possible into the caliper bores to provide
the clearance necessary to accommodate the unworn pads. It should be
possible to do this with hand pressure alone. If any undue stiffness is
encountered the caliper assembly should be dismantled for examination
as described in Chapter 6. While pushing the pistons back, maintain a
careful watch on the fluid level in the handlebar reservoir. If the reservoir
has been overfilled, the surplus fluid will prevent the pistons returning
fully and must be removed by soaking it up with a clean cloth. Take care
to prevent fluid spillage. Apply a thin smear of caliper grease to the outer
edge and rear surface of the moving pad and to the pad retaining pins.
Take care to apply caliper grease to the metal backing of the pad only and
not to allow any grease to contaminate the friction material. Carefully fit
the pad anti-rattle spring, ensuring that it is correctly located, and insert
the moving pad into its aperture in the caliper mounting bracket. Check
that the pad is free to slide in the mounting bracket then refit the second
pad and insert the two pad retaining pins, tightening them securely to
the specified torque setting. On MTX125/200 models, do not forget to
place the anti-squeal shim and retainer correctly in position as the pads
are refitted. On MBX125 models, note that the pads must be refitted so

that the larger areas of friction material are downwards when the caliper
is in place on the machine.

Replace the caliper assembly on the machine ensuring that the pads
engage correctly on the disc and that the hose is not twisted, then refit
the mounting bolts and tighten them securely, to the specified torque
setting if possible. Check that the pad pins are securely tightened and
refit the threaded plugs. Apply the brake lever gently and repeatedly to
bring the pads firmly into contact with the disc until full brake pressure is
restored. Be careful to watch the fluid level in the reservoir; if the pads
have been re-used it will suffice to keep the level above the lower level
mark, by topping-up if necessary, but if new pads have been fitted the
level must be restored to the upper level line described above by topping
up or removing surplus fluid as necessary. Refit the reservoir cover,
gasket (where fitted) and diaphragm as described above.

Before taking the machine out on the road, be careful to check for
fluid leaks from the system, and that the front brake is working correctly.
Remember also that new pads, and to a lesser extent, cleaned pads will
require a bedding-in period before they will function at peak efficiency.
Where new pads are fitted use the brake gently but firmly for the first 50 –
100 miles to enable the pads to bed in fully.

Drum brakes will require regular adjustment to compensate for shoe
wear. At the front (MTX125/200 RW-D models), the brake is adjusted at
the cable lower end, on the brake backplate; reserve the handlebar
adjuster for quick roadside adjustments and ensure it is screwed fully in
before using the lower adjuster. Note also that the clamp securing the
cable to the fork lower leg must be released so that the cable can move.
On the rear brake (all models) adjustment is made by rotating the nut at
the rear end of the operating rod.

Support the machine so that the wheel is clear of the ground, spin the
wheel and tighten the adjusting nut until a rubbing sound is heard as the
shoes begin to contact the drum, then slacken the nut by one or two
turns until the sound ceases. Tighten the adjuster locknut (front brakes
only) then spin the wheel and apply the brake hard to settle the
components. Check that the adjustment has not altered and tighten
securely all disturbed fasteners. This should approximate the specified
setting which is that there should be 25 – 35 mm (1 – 1.4 in) of free play
before the brake begins to engage the drum, measured at the handlebar
lever ball end or brake pedal tip, as appropriate. On rear brakes only,
switch on the ignition and check that the stop lamp lights just as all free
play has been taken up and the brake is beginning to engage. This is
adjusted by holding steady the stop lamp rear switch body and rotating
the plastic sleeve nut to raise or lower the switch as necessary; do not
allow the body to rotate or its terminal will be damaged. Check the
switch setting whenever the rear brake is adjusted.

Complete brake maintenance by checking that the wheels are free to
rotate easily, and then lubricate all exposed lengths of inner cable, all
lever or linkage pivots, and the stoplamp switch. To prevent the risk of oil
finding its way on to the tyres or the brake friction material do not oil
excessively the brake components; a few drops of oil at each point will
suffice.

Push pistons back to make room for new pads – note anti-rattle
spring

MBX125 – note correct fitted position of brake pads

Front brake adjustment – MTX125/200 RW-D – slacken cable clamp on fork lower leg ...

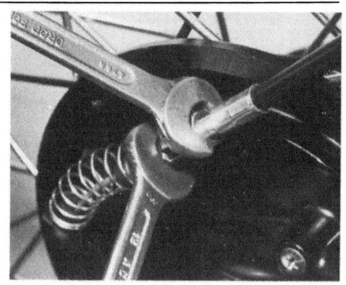

... and adjust brake as shown

Adjusting rear brake – all models

Do not allow switch body to rotate when adjusting switch setting

Check the condition of the brake operating cable (where appropriate) and linkage, ensuring that all components are in good condition, properly lubricated and securely fastened. Dismantle the rear brake pedal pivot and grease it at regular intervals. Note that the operating mechanism is at its most efficient when, with the brake correctly adjusted and applied fully, the angle between the cable or rod and the operating arm on the brake backplate does not exceed 90°. This can be adjusted by removing the operating arm from the brake camshaft and rotating it by one or two splines until the angle is correct. Ensure that all components are correctly secured on reassembly.

Drum brake shoe wear is checked by applying the brake firmly and looking at the wear indicator marks on the backplate. If the indicator pointer on the camshaft is aligned with, or has moved beyond, the fixed index mark cast on the backplate, the shoes are worn out and must be renewed as a pair. This involves the removal of the wheel from the machine as described in Chapter 6 so that the brake components can then be dismantled, cleaned, checked for wear, and reassembled following the instructions given in the same Chapter. It is important that moving parts such as the brake camshaft are lubricated with a smear of high melting-point grease on reassembly.

If pointer on camshaft aligns with cast arrow on backplate, brake shoes must be renewed

8 Check the suspension and steering

Support the machine so that it is secure with the front wheel clear of the ground, then grasp the front fork legs near the wheel spindle and push and pull firmly in a fore and aft direction. If play is evident between the top and bottom fork yokes and the steering head, the steering head bearings are in need of adjustment. Imprecise handling or a tendency for the front forks to judder may be caused by this fault.

Bearing adjustment is correct when the adjuster ring is tightened, until resistance to movement is felt and then loosened 1/8 to 1/4 a turn. The adjuster ring should be rotated by means of a C-spanner after slackening the steering stem nut.

Take great care not to overtighten the adjuster ring. It is possible to place a pressure of several tons on the head bearings by over-tightening even though the handlebars may seem to turn quite freely. Overtight bearings will cause the machine to roll at low speeds and give imprecise steering. Adjustment is correct if there is no play in the bearings and the handlebars swing to full lock either side when the machine is supported with the front wheel clear of the ground. Only a light tap on each end should cause the handlebars to swing. Secure the adjuster ring by tightening the steering stem top nut to the specified torque setting, then check that the setting has not altered.

At the same time as the steering head bearings are checked, take the opportunity to examine closely the front and rear suspension. Ensure that the front forks work smoothly and progressively by pumping them up and down whilst the front brake is held on. Any faults revealed by this check should be investigated further, as any deterioration in the stability of the machine can have serious consequences. Check carefully for signs of leaks around the front fork oil seals. If any damage is found, it must be repaired immediately as described in the relevant Sections of Chapter 5. On MBX125 models inspect the stanchions, looking for signs of chips or other damage, then lift the dust excluder at the top of each fork lower leg and wipe away any dirt from its sealing lips or above the fork oil seal. Pack grease above the seal and refit the dust excluder. Note that none of this would be necessary, and fork stiction would be reduced, if gaiters are fitted; they are available from any good motorcycle dealer. For obvious reasons this is not necessary on MTX125/200 models but it is still advisable to lift the gaiters at regular intervals to check the seals; smear grease over the stanchion to prevent corrosion.

To check the swinging arm support the machine so that the rear wheel is clear of the ground then pull and push horizontally at the rear end of the swinging arm; there should be no discernible play at the pivot. If the rear suspension is not to be dismantled for lubrication, pump grease into the swinging arm pivot via the grease nipple provided until fresh grease is seen at both ends of the pivot. Honda recommend the use of a molybdenum disulphide-based grease.

The rear suspension linkage consists of several highly-stressed bearing surfaces which are not fitted with grease nipples (except in the case of later MTX125/200 models) and which are very exposed to all the water, dirt and salt thrown up by the rear wheel. Regular cleaning and greasing is essential; owners of MBX125 and MTX125/200 RW-D models should note that this is greatly simplifed if the linkage were dismantled for all bearings to be fitted with grease nipples. A local Honda Service Agent should be able to tell you of someone competent to undertake such work. Note that not only are the linkage components expensive to renew if allowed to wear out through lack of attention, but the machine will almost certainly fail its DOT certificate test if such wear is found.

To check the linkage components, two people are required; one to sit on the machine and bounce the rear suspension while the other watches closely the action of the various components. If any squeaks or other noises are heard, if the linkage appears stiff, or if any signs of dry bearings or other wear or damage are detected, the rear suspension should be removed as a complete assembly and dismantled for thorough cleaning, checking and greasing. Where grease nipples are fitted, remove the weight from the suspension by supporting the machine so that the rear wheel is clear of the ground, then inject grease into each bearing until all old grease is expelled and new grease can be seen issuing from each end of the bearing. If nipples are not fitted, the linkage must be dismantled so that the bearings can be cleaned and packed with new grease. Refer to Chapter 5.

9 Lubricate the cables, stands and controls

At regular intervals, all control and instrument drive cables, all control pivots and stands should be checked for wear or damage and lubricated, dismantling them where necessary and removing all traces of dirt or corrosion. This operation must be carried out to prevent excessive wear and to ensure that the various components can be operated smoothly and easily, in the interests of safety. The opportunity should be taken to examine closely each component, renewing any that show signs of excessive wear or of any damage.

The twistgrip is removed by unscrewing the screws which fasten both halves of the handlebar right-hand switch assembly. The throttle cable upper end nipple can be detached from the twistgrip with a suitable pair of pliers and the twistgrip slid off the handlebar end. Carefully clean and examine the handlebar end, the internal surface of the twistgrip, and the two halves of the switch cluster. Remove any rough burrs with a fine file, and apply a coating of grease to all the bearing surfaces. Slide the twistgrip back over the handlebar end, insert the throttle cable end nipple into the twistgrip flange, and reassemble the switch cluster. Check that the twistgrip rotates easily and that the throttle snaps shut as soon as it is released. Tighten the switch retaining screws securely, but do not overtighten them.

Although the regular daily checks will ensure that the control cables are lubricated and maintained in good order, it is recommended that a positive check is made on each cable at this mileage/time interval to ensure that any faults will not develop unnoticed to the point where smooth and safe control operation is impaired. If any doubt exists about the condition of any of the cables, the component in question should be removed from the machine for close examination. Check the outer cables for signs of damage, then examine the exposed portions of the inner cables. Any signs of kinking or fraying will indicate that renewal is required. To obtain maximum life and reliability from the cables they should be thoroughly lubricated using light machine oil. To do the job properly and quickly use one of the hydraulic cable oilers available from most motorcycle shops. Free one end of the cable and assemble the cable oiler as described by the manufacturer's instructions. Operate the oiler until oil emerges from the lower end, indicating that the cable is lubricated throughout its length. This process will expel any dirt or moisture and will prevent its subsequent ingress.

If a cable oiler is not available, an alternative is to remove the cable from the machine. Hang the cable upright and make up a small funnel arrangement using plasticine or by taping a plastic bag around the upper

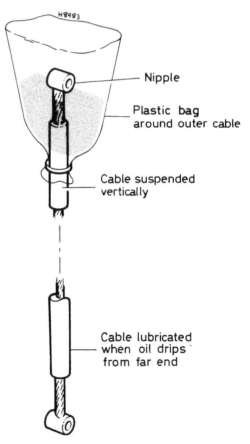

Oiling a control cable

end. Fill the funnel with oil and leave it overnight to drain through. Note that where nylon-lined cables are fitted, they should be used dry or lubricated with a silicone-based lubricant suitable for this application. On no account use ordinary engine oil because this will cause the liner to swell, pinching the cable. The throttle/oil pump cable junction box should be dismantled, cleaned and reassembled using a 'dry' graphite-based lubricant or any lightweight grease or aerosol spray lubricant. Do not use ordinary grease as this will at best make the throttle action heavy; at worst it could cause an accident by preventing the smooth return of the throttle.

Each instrument drive cable is secured at its upper mounting by a knurled, threaded ring and at its bottom mounting by a single retaining screw in the speedometer drive gearbox, front brake backplate, or crankcase right-hand cover, as appropriate. Remove the cable by using a pair of pliers to unscrew the knurled ring, and a screwdriver to slacken and remove the retaining screw. It should be possible to withdraw the cable carefully from its mountings and remove it from the machine. Remove the inner cable by pulling it out from the bottom of the outer. Carefully examine the inner cable for signs of fraying, kinking, or for any shiny areas which will indicate tight spots, and the outer cable for signs of cracking, kinking or any other damage. Renew either cable if necessary. To lubricate the cable, smear a small quantity of grease on to the lower length only of the inner. Do not allow any grease on the top six inches of the cable as the grease will work its way rapidly up the length of the cable as it rotates and get into the instrument itself. This will rapidly ruin the instrument which will then have to be renewed. Insert the inner cable in the outer and refit the cable.

Check all pivots and control levers, cleaning and lubricating them to prevent wear or corrosion. Where necessary, dismantle and clean any moving part which may have become stiff in operation. Similarly clean, check and grease the stand pivots and ensure that the return spring(s) hold the stand(s) securely. On later MTX125/200 models check the condition of the side stand rubber pad; if this is worn at any point to or beyond the arrowed wear limit line it must be renewed using only a pad marked '260 lb/117 kg or more'.

When refitting the cables onto the machine, ensure that they are routed in easy curves and that full use is made of any guide or clamps that have been provided to secure the cable out of harm's way. Adjustment of the individual cables is described under other Routine Maintenance tasks.

Be very careful to ensure that all controls are correctly adjusted and are functioning correctly before taking the machine out on the road.

10 Check all fittings, fasteners, lights and signals

Check around the machine, looking for loose nuts, bolts or screws, retightening them as necessary.

It is advisable to lubricate the handlebar switches and stop lamp switches with WD40 or a similar water dispersant lubricant. This will keep the switches working properly and prolong their life, especially if the machine is used in adverse weather conditions. Check that all lights, turn signals, the horn and speedometer are working properly and that their mountings and connections are securely fastened.

Eight monthly, or every 5000 miles (8000 km)

Repeat all tasks listed under the previous mileage/time headings, then carry out the following:

1 Decarbonise the engine and exhaust system
Cylinder head and barrel

Refer to Chapter 1 and remove the cylinder head and barrel. It is necessary to remove all carbon from the head, barrel and piston crown whilst avoiding removal of the metal surfaces on which it is deposited. At the same time check the ATAC chamber passages (all models) and clean the valve (MTX200). Refer to Section 12 of Chapter 3. Take care when dealing with the soft alloy head and piston. Never use a steel scraper or screwdriver. A hardwood, brass or aluminium scraper is ideal as these are harder than the carbon but not harder than the underlying metal. With the bulk of carbon removed, use a brass wire brush. Finish the head and piston with metal polish; a polished surface will slow the subsequent build-up of carbon. Clean out the cylinder barrel ports to prevent the restriction of gas flow. Remove all debris by washing each component in paraffin whilst observing the necessary fire precautions. Renew the piston rings, if necessary, on reassembly.

Exhaust system

Cleaning the exhaust system is complicated by the fact that the baffles cannot be removed for cleaning, although on MTX125/200 models a plate secured to the tailpipe/silencer underside by two bolts can be removed for cleaning. Honda recommend that the carbon deposits are loosened from inside the tailpipe and that the engine is then started with a rag blocking the tailpipe so that the loose deposits are blown out through the slot in the bottom of the pipe. This can at best be described as hopeful. MBX125 owners should note that the baffles appear to be retained by three rivets; it may be possible for the more enterprising owner to drill out these rivets, extract the baffles for thorough cleaning, and then devise a method of retaining the baffles (perhaps by fitting retaining screws) so that they are held securely but can be removed when necessary.

It is necessary to check the exhaust system at regular intervals so that the build-up of carbon deposits and of oily sludge can be dealt with as soon as possible.

First check that the exhaust system is securely fastened and that there are no exhaust leaks. Renew the exhaust port gasket and, on MTX125/200 models only, the rubber seal between the two parts of the system, if leaks are found at either of these points. Some idea of the condition of the exhaust can be gained by looking at the tailpipe. If the mixture and oil pump settings are correct, and the machine is ridden normally, there should be a thin film of sooty black carbon with a slight trace of oiliness. If the machine is ridden hard, the deposits will tend to

MTX125/200 models are fitted with removable plate in tailpipe underside for decoking

MBX125 models – exhaust baffles are not removable

be a lighter colour, almost grey, and there will be no traces of oil. If any of the above is found, and the spark plug electrode colour, carburettor settings, air filter condition, and the oil pump settings are known to be satisfactory from the previous Routine Maintenance operations, then the carbon deposits inside the exhaust system will be kept to a minimum and will take a long time to build up to the point where a full decarbonising operation is necessary. On the other hand, if the rear of the exhaust is excessively oily and the carbon deposits rather thicker than those described above, then either one of the settings mentioned above is incorrect, causing the engine to run inefficiently so that it produces too much waste in the form of carbon, or the machine is being ridden too slowly which produces the same symptoms. If the engine settings are known to be satisfactory thanks to the previous Routine Maintenance operations and the carbon build-up is caught at an early stage, then the simplest and most satisfying method of cleaning the exhaust is to take the machine on a good hard run until the exhaust is too hot to touch and the excessive smoke produced by the dispersal of the carbon/oil build-up has disappeared. This is a well known trick employed by many mechanics to restore lost power, and can have quite dramatic results. If the build-up has been allowed to develop too much, however, the complete exhaust must be removed from the machine and decarbonised using one of the methods described below.

Three possible methods of cleaning the system exist, the first, which will only be really effective if the deposits are very oily, consists of flushing the system with a petrol/paraffin mixture. It will be evident that great care must be taken to prevent the risk of fire when using this method, and that the system must be hung overnight in such a way that all the mixture will drain away. It will be necessary to use a suitable scraping tool to remove any hardened deposits from the front length of the exhaust pipe and from the exhaust port of the cylinder barrel.

The second method will involve the use of a welding torch or powerful blow lamp to burn off the deposits. This usually results in the production of a great deal of smoke and fumes and so must be carried out in a well-ventilated area. It also requires some considerable skill in the use of welding equipment if it is to be fully effective, and if personal injury or damage to the exhaust system is to be avoided. Another drawback is that the excessive heat created will destroy the painted finish of the system, and that repainting will be necessary. With severely blocked systems, it may even be necessary to cut the exhaust open so that the flame can be applied to the blocked section, and the exhaust welded up afterwards. In short this method, while quite effective, should only be employed by a person with welding equipment and the necessary degree of skill in its use. Note also that Honda specifically advise against the use of this method. The best solution would be for the owner to remove the exhaust from the machine and to take it to a local dealer or similar expert for the work to be carried out.

The third possible method of exhaust cleaning is to use a solution such as caustic soda to dissolve the deposits. While this is a lengthy and time-consuming operation, it is the simplest and most effective method that can be used by the average owner, and is therefore described in detail in the following paragraphs. To clean the silencer casing and exhaust pipe assembly, remove the system from the machine and suspend it from its silencer end. Block up the end of the exhaust pipe with a cork or wooden bung. If wood is used, allow an outside projection of three or four inches with which to grasp the bung for removal.

The mixture used is a ratio of 3 lbs caustic soda to a gallon of fresh water. This is the strongest solution ever likely to be required. Obviously the weaker the mixture the longer the time required for the carbon to be dissolved. Note, whilst mixing the solution, that the caustic soda should be added to the water gradually, whilst stirring. Never pour water into a container of caustic soda powder or crystals; this will cause a violent reaction to take place which will result in great danger to one's person.

Commence the cleaning operation by pouring the solution into the system until it is quite full. Do not plug the open end of the system. The solution should now be left overnight for its dissolving action to take place. Note that the solution will continue to give off noxious fumes throughout its dissolving process; the system must therefore be placed in a well ventilated area. After the required time has passed, carefully pour out the solution and flush the system through with clean, fresh water. The cleaning operation is now complete.

Bear in mind that it is very important to take great care when using caustic soda as it is a very dangerous chemical. Always wear protective clothing, this must include proper eye protection. If the solution does come into contact with the eyes or skin it must be washed clear immediately with clean, fresh, running water. In the case of an eye

becoming contaminated, seek expert medical advice immediately. Also, the solution must not be allowed to come into contact with aluminium alloy – especially at the above recommended strength – caustic soda reacts violently with aluminium and will cause severe damage to the component.

2 Check the cooling system

The cooling system should be checked for signs of leakage or damage and any suspect hoses renewed; this is best carried out in conjunction with decarbonisation, when most of the components will have been disturbed. The used coolant can be put back into the system if it is clean, but note that it must be renewed every two years as a matter of course. Refer to Chapter 2 for full instructions.

3 Renew the spark plug

The spark plug should be renewed at this interval, regardless of its apparent condition, as it will have passed peak efficiency. Check that the new plug is of the correct type and heat range and that it is gapped correctly before it is fitted.

4 Check the wheels, bearings and tyres

To check the wheels, support the machine securely so that the wheel to be checked is clear of the ground and can spin freely. If necessary slacken the brake adjustment and in the case of the rear wheel, disconnect the final drive chain. Proceed as follows:

MBX125 models

Spin the wheel and check for rim alignment by placing a pointer close to the rim edge. If the total radial or axial alignment variation is greater than 2.0 mm (0.08 in) the manufacturer recommends that the wheel is renewed. This policy is, however, a counsel of perfection and in practice a larger runout may not affect the handling properties excessively.

Although Honda do not offer any form of wheel rebuilding facility, a number of private engineering firms offer this service. It should be noted however, that Honda do not approve of this course of action.

Check the rim for localised damage in the form of dents or cracks. The existence of even a small crack renders the wheel unfit for further use unless it is found that a permanent repair is possible using arc-welding. This method of repair is highly specialised and therefore the advice of a wheel repair specialist should be sought.

Inspect the spoke blades for cracking and security. Check carefully the area immediately around the rivets which pass through the spokes and into the rim. In certain circumstances corrosion may occur between the spokes, rivets and rim due to the use of different metals.

MTX125/200 models

Examine the rim for serious corrosion or impact damage. Slight deformities can often be corrected by adjusting spoke tension. Serious damage and corrosion will necessitate renewal, which is best left to an expert.

Place a wire pointer close to the rim and rotate the wheel to check it for runout. If the rim is more than 2.0 mm (0.08 in) out of true in the radial or axial planes, check spoke tension by tapping them with a screwdriver. A loose spoke will sound quite different from those around it. Worn bearings will also cause rim runout.

Adjust spoke tension by turning the square-headed nipples with the appropriate spoke key which can be purchased from a dealer. With the spokes evenly tensioned, remaining distortion can be pulled out by tightening the spokes on one side of the wheel and slackening those directly opposite. This will pull the rim across whilst maintaining spoke tension.

More than slight adjustment will cause the spoke ends to protrude through the nipple and chafe the inner tube, causing a puncture. Remove the tyre and tube and file off the protruding ends. The rim band protects the tube against chafing; check it is in good condition before fitting.

Check spoke tension and general wheel condition regularly. Frequent cleaning will help prevent corrosion. Replace a broken spoke immediately because the load taken by it will be transferred to adjacent spokes which may fail in turn.

All models

To check the wheel bearings, grasp each wheel firmly at the top and bottom and attempt to rock it from side to side; any free play indicates worn bearings which must be renewed as described in Chapter 6. Make a careful check of the tyres, looking for signs of damage to the tread or sidewalls and removing any embedded stones etc. Renew the tyre if the tread is excessively worn or if it is damaged in any way.

Annually, or every 7500 miles (12 000 km)

The annual maintenance operation should be regarded as a minor overhaul. Carry out all the applicable tasks listed in the previous mileage/time service headings, then complete the following:

1 Change the transmission oil

Oil changes are much quicker and more efficient if the machine is taken for a journey long enough to warm the oil up to normal operating temperature so that it is thin and is holding any impurities in suspension. Position a container of at least 700 cc (1.2 pint) capacity beneath the engine. Remove the drain plug from the underside of the engine unit. Whilst waiting for the oil to drain, examine the plug sealing washer and renew if damaged.

On completion of draining, refit the plug and tighten it to the recommended torque setting of 2.0 - 2.5 kgf m (14.5 - 18 lbf ft). Remove the filler plug and fill the gearbox with 500 cc (0.9 pint) of SAE 10W/30 SE or SF engine oil. Check the oil level as described under the four monthly/2500 mile heading and refit the filler plug. Check carefully for any signs of oil leaks.

2 Clean the engine oil tank filter

If suitable care is taken in adding only clean oil to the tank, this task should not be necessary. In the event that it is, proceed as follows.

First remove the oil pump cover and disconnect the tank/pump feed pipe from the pump. Allow the tank to drain completely into a clean container. On MBX125 models remove the side panels and the seat, disconnect and remove the battery, then remove the battery tray which is retained by four bolts; the lowest of these also retains the coolant expansion tank which can be put to one side without disconnecting it. Remove the single oil tank mounting bolt, disconnect the oil level switch wires and manoeuvre the tank clean of the machine taking care that the oil feed pipe is released from any retaining clamps or ties. On MTX125/200 models the tank filter can be reached while the tank is in place; it is merely necessary to pack rag under the tank union to soak up any spilt oil.

Slacken the screw clamping the union to the tank and pull out the union and filter gauze. Wipe any surplus oil and dirt from the tank and clean the gauze by blowing through it with compressed air or by flushing it in clean petrol, taking care to prevent the risk of fire. Thoroughly clean all other components so that no dirt is introduced on reassembly. Refit the gauze to the union and fit the filter assembly into the tank, tightening securely the screw clamp.

On MBX125 models refit the tank to the machine tightening all bolts to their specified torque settings and connecting the switch wires. Check that the feed pipe is routed correctly to the pump and secured by any clamps or ties provided. On all models, having checked that it is clean, pour the drained oil back into the tank and allow a small quantity to drip out of the feed pipe before connecting it again to the oil pump; this will speed up the bleeding procedure which must now be carried out as described in Chapter 3 to ensure that there are no air bubbles in the system. Finally, check that there are no oil leaks and wash off any spilt oil, check that all disturbed components (particularly the feed pipe) are correctly refitted and securely fastened, refit the oil pump cover and top up the tank to the bottom of the filler neck.

3 Renew the brake fluid

It is necessary to renew the brake fluid at this interval to preserve maximum brake efficiency by ensuring that the fluid has not been contaminated and deteriorated to an unsafe degree.

Before starting work, obtain a new, full can of SAE J1703 or DOT3 hydraulic fluid and read carefully the Section on brake bleeding in Chapter 6. Prepare the clear plastic tube and glass jar in the same way as for bleeding the hydraulic system, open the bleed nipple by unscrewing it $1/4 - 1/2$ a turn with a spanner and apply the front brake lever gently and repeatedly. This will pump out the old fluid. **Keep the master cylinder reservoir topped up at all times,** otherwise air may enter the system and greatly lengthen the operation. The old brake fluid is invariably much darker in colour than the new, making it easier to see when it is pumped out and the new fluid has completely replaced it.

When the new fluid appears in the clear plastic tubing completely uncontaminated by traces of old fluid, close the bleed nipple, remove the plastic tubing and replace the rubber cap on the nipple. Top the master cylinder reservoir up to above the lower level mark, unless the brake pads have been renewed in which case the reservoir should be topped up to its higher level. Clean and dry the rubber diaphragm, fold it into its compressed state and refit the diaphragm and reservoir cover, tightening securely the retaining screws.

Wash off any surplus fluid and check that the brake is operating correctly before taking the machine out on the road.

4 Grease the steering head bearings

Referring to the relevant Sections of Chapter 5, dismantle the front forks and steering head, clean all components and check them for wear, renewing any worn items. Reassemble, packing the steering head bearings with fresh grease.

Transmission oil drain plug is situated in crankcase underside

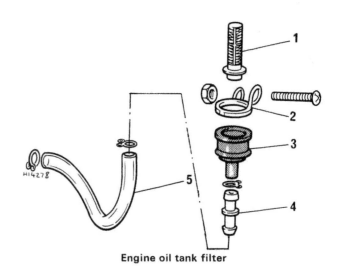

Engine oil tank filter

1 Filter gauze   4 Joint
2 Clamp   5 Feed pipe
3 Union

5 Change the front fork oil

The fork oil must be changed at regular intervals to prevent the inevitable reduction in fork performance which results as the oil deteriorates in service. This task can be carried out as part of the previous operation and is outlined below. Refer to the relevant part of Chapter 5 for full information.

Place a sheet of cardboard against the wheel to keep oil off the brake or tyre, place a suitable container under the fork leg and remove the drain plug. Depress the forks several times to expel as much oil as possible, then repeat the process on the remaining leg. Leave the machine for a few minutes to allow any residual oil to drain to the bottom, then pump the forks again to remove it.

Renewing their sealing washers if worn or damaged, refit and tighten securely the drain plugs, then remove the fork leg top plugs, having first depressed the valve cores to release the air pressure. Withdraw the fork spring(s) noting carefully which way up they are fitted.

When adding fork oil, note that while the standard recommended oil is ATF or Automatic Transmission Fluid, the fork action can be varied to suit the rider's needs by using different grades of oil; the lighter the oil, the lighter the damping effect. Proprietary fork oils are available in several grades from most good motorcycle dealers. Add all but the last 10 – 20 cc of oil to each leg then slowly pump the legs up and down to expel as much air as possible and to distribute the oil fully throughout the leg. Support the machine so that it is securely upright with both legs fully compressed. The level of oil in each leg must now be set.

The level is measured from the top of the stanchion to the top of the oil when the fork top plugs and springs are removed and the forks are fully compressed; a dipstick of suitable length can be made from welding rod or similar. It is essential that the level is at exactly the correct height and that the quantity in both legs is precisely the same; add or remove oil to achieve this. Note that the amount of fork oil in each leg is relatively unimportant; the oil level is the essential factor.

When the level is correct, refit the spring(s) and top plugs as described in Chapter 5, set the required air pressure as described earlier in Routine Maintenance and check that the forks move smoothly and correctly throughout their full travel before using the machine on the road.

6 Dismantle, clean and grease the rear suspension

Regardless of whether grease nipples are fitted or not, it is advisable to remove the rear suspension from the machine so that it can be dismantled, cleaned, checked for wear and reassembled as described in Chapter 5. Pack each bearing with grease on rebuilding and renew any damaged, or worn components.

7 Grease wheel bearings, speedometer drive and brake camshafts

To prolong their life as much as possible, the front and rear wheels should be removed from the machine so that their bearings can be driven out, cleaned, checked for wear and renewed if necessary, then repacked with fresh high-melting point grease and refitted. At the same time the speedometer drive should be cleaned and packed with grease, and on drum brakes only, the brake camshaft should be removed, cleaned and greased. Refer to the relevant Sections of Chapter 6.

Additional routine maintenance

1 Overhaul the hydraulic brake components

At intervals of every two years, the master cylinder and caliper of the hydraulic front brake should be dismantled and checked. This is because their seals will deteriorate with age whether the machine is ridden a great deal or hardly at all. The seals should be renewed regardless of their apparent condition and the master cylinder and caliper bores, as well as the caliper pistons, should be checked very carefully for wear or damage. The brake fluid will be changed as part of the annual/7500 mile service. Check the brake hose very carefully indeed and renew it if any signs of splitting, cracking or chafing are seen. While Honda make no specific recommendation to this effect, it is usual to renew the hoses at the same time as the seals and for the same reason. Refer to Chapter 6 for full information.

2 Renew the coolant

To minimise the build-up of deposits in the cooling system and to ensure maximum protection against its freezing, the coolant should be drained completely, the system should be flushed out and checked for leaks or damage and new coolant mixed for refilling. This should be done every two years, and is described in Chapter 2.

3 Cleaning the machine

Keeping the motorcycle clean should be considered as an important part of the routine maintenance, to be carried out whenever the need arises. A machine cleaned regularly will not only succumb less speedily to the inevitable corrosion of external surfaces, and hence maintain its market value, but will be far more approachable when the time comes for maintenance or service work. Furthermore, loose or failing components are more readily spotted when not partially obscured by a mantle of road grime and oil.

Surface dirt should be removed using a sponge and warm, soapy water; the latter being applied copiously to remove the particles of grit which might otherwise cause damage to the paintwork and polished surfaces.

Oil and grease is removed most easily by the application of a cleaning solvent such as 'Gunk' or 'Jizer'. The solvent should be applied when the parts are still dry and worked in with a stiff brush. Large quantities of water should be used when rinsing off, taking care that water does not enter the carburettors, air cleaners or electrics.

Application of a wax polish to the cycle parts and a good chrome cleaner to the chrome parts will give a good finish. Always wipe the machine down if used in the wet, and make sure the chain is well oiled. There is less chance of water getting into control cables if they are regularly lubricated, which will prevent stiffness of action.

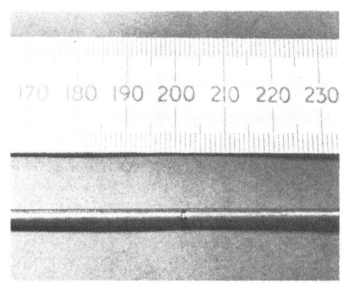

Dipstick should be made from welding rod or similar ...

... and used to check fork oil level on refitting forks

Chapter 1 Engine, clutch and gearbox

Contents

Specifications

Unless otherwise stated, information applies to all models

Engine

	125 models	**200 models**
Type	Two stroke, single cylinder, water-cooled	
Capacity	124.6 cc (7.6 cu in)	193.9 cc (11.8 cu in)
Bore	56.0 mm (2.21 in)	67.0 mm (2.64 in)
Stroke	50.6 mm (1.99 in)	55.0 mm (2.17 in)
Compression ratio	7.5:1	
Compression pressure – at cranking speed, engine fully warmed up:		
MBX125	12.5 kg/cm² (178 psi)	
MTX125/200	13.0 ± 2.0 kg/cm² (185 ± 28 psi)	

Cylinder head

Type	Cast aluminium alloy, with separate water jacket	
Gasket face maximum warpage	0.10 (0.0039 in)	

Cylinder barrel

	125 models	**200 models**
Standard bore ID	56.003 – 56.018 mm	67.003 – 67.023 mm
	(2.2048 – 2.2054 in)	(2.6379 – 2.6387 in)
Service limit	56.080 mm (2.2079 in)	67.080 mm (2.6409 in)
Piston/cylinder clearance	0.033 – 0.063 mm	0.033 – 0.068 mm
	(0.0013 – 0.0025 in)	(0.0013 – 0.0027 in)
Service limit	0.160 mm (0.0063 in)	0.160 mm (0.0063 in)

Piston rings
Top:
Type .... Keystone
Top surface marking .... N, R or T
End gap – installed .... 0.15 – 0.30 mm (0.0059 – 0.0118 in)
Service limit .... 0.35 mm (0.0138 in)
Ring/ring groove clearance .... N/App
Second:
Type .... Plain, with expander
Top surface marking .... N, R or T
End gap – installed .... 0.15 – 0.30 mm (0.0059 – 0.0018 in)
Service limit .... 0.35 mm (0.0138 in)
Ring/ring groove clearance .... 0.015 – 0.045 mm (0.0006 – 0.0018 in)
Service limit .... 0.100 mm (0.0039 in)

Piston

	125 models	200 models
Standard OD	55.955 – 55.970 mm (2.2030 – 2.2035 in)	66.955 – 66.970 mm (2.6360 – 2.6366 in)
Service limit	55.920 mm (2.2016 in)	66.920 mm (2.6346 in)
Gudgeon pin bore ID	16.002 – 16.008 mm (0.6300 – 0.6302 in)	16.002 – 16.008 mm (0.6300 – 0.6302 in)
Service limit	16.030 mm (0.6311 in)	16.030 mm (0.6311 in)
Gudgeon pin OD	15.994 – 16.000 mm (0.6297 – 0.6299 in)	15.994 – 16.000 mm (0.6297 – 0.6299 in)
Service limit	15.980 mm (0.6291 in)	15.980 mm (0.6291 in)
Piston/gudgeon pin clearance	0.002 – 0.014 mm (0.00008 – 0.00055 in)	0.002 – 0.014 mm (0.00008 – 0.00055 in)
Service limit	0.050 mm (0.0020 in)	0.050 mm (0.0020 in)

Crankshaft

	125 models	200 models
Connecting rod small end ID	20.005 – 20.017 mm (0.7876 – 0.7881 in)	21.005 – 21.017 mm (0.8270 – 0.8274 in)
Service limit	20.030 mm (0.7886 in)	21.030 mm (0.8280 in)
Big-end side clearance	0.15 – 0.55 mm (0.0059 – 0.0217 in)	0.15 – 0.55 mm (0.0059 – 0.0217 in)
Service limit	0.85 mm (0.0335 in)	0.85 mm (0.0335 in)
Big-end maximum radial free play	0.05 mm (0.0020 in)	0.05 mm (0.0020 in)
Maximum runout at journals	0.10 mm (0.0039 in)	0.10 mm (0.0039 in)

Primary drive
Type .... Gear
Reduction ratio 3.050:1 (61/20T)

Clutch

	125 models and MTX200 RW-F	MTX200 RW-D
Type	Wet, multi-plate	
Number of friction plates	5	6
Number of plain plates	4	5
Number of springs	4	
Friction plate thickness	2.9 – 3.0 mm (0.1142 – 0.1181 in)	
Service limit	2.5 mm (0.0984 in)	
Plain plate maximum warpage	0.2 mm (0.0079 in)	
Spring free length	34.8 mm (1.3701 in)	
Service limit	33.8 mm (1.3307 in)	
Spring preload	20.8 – 22.4 kg at 23.2 mm (45.9 – 49.4 lb at 0.91 in)	25.1 – 26.7 kg at 20.9 mm (55.3 – 58.9 lb at 0.82 in)
Outer drum centre ID	23.000 – 23.021 mm (0.9055 – 0.9603 in)	
Service limit	23.060 mm (0.9079 in)	
Outer drum bearing sleeve dimensions:		
Standard OD	22.930 – 22.950 mm (0.9028 – 0.9035 in)	
Service limit	22.800 mm (0.8976 in)	
Standard ID	16.988 – 17.010 mm (0.6688 – 0.6697 in)	
Service limit	17.040 mm (0.6709 in)	

Balancer/water pump drive idler shaft

	Standard	Service limit
OD at right-hand bearing:		
MBX125, early MTX125/200 RW-D models	11.972 – 11.987 mm (0.4713 – 0.4719 in)	11.940 mm (0.4701 in)
Later MTX125/200 models	Approx 14 mm (0.55 in)	N/Av
OD at water pump drive gear/crankcase breather – all models	9.972 – 9.983 mm (0.3926 – 0.3930 in)	9.930 mm (0.3909 in)

Kickstart mechanism
Pinion ID .... 20.020 – 20.041 mm (0.7882 – 0.7890 in)
Service limit .... 20.100 mm (0.7913 in)
Shaft OD – at pinion bearing surface .... 19.959 – 19.980 mm (0.7858 – 0.7866 in)
Service limit .... 19.920 mm (0.7843 in)
Idler gear ID .... 20.020 – 20.041 mm (0.7882 – 0.7890 in)
Service limit .... 20.100 mm (0.7913 in)

Idler gear bush dimensions:
    Standard OD .... 19.984 – 19.995 mm (0.7868 – 0.7872)
    Service limit .... 19.900 mm (0.7835 in)
    Standard ID .... 17.016 – 17.034 mm (0.6699 – 0.6706 in)
    Service limit .... 17.100 mm (0.6732 in)
Idler gear (output) shaft OD .... 16.966 – 16.984 mm (0.6680 – 0.6687 in)
Service limit .... 16.940 mm (0.6669 in)

Gearbox

	MBX125	MTX125	MTX200
Type	6-speed constant mesh		

Reduction ratios:

	MBX125	MTX125	MTX200
1st	2.833:1 (34/12T)	3.091:1 (34/11T)	3.091:1 (34/11T)
2nd	1.875:1 (30/16T)	2.000:1 (30/15T)	1.938:1 (31/16T)
3rd	1.421:1 (27/19T)	1.474:1 (28/19T)	1.368:1 (26/19T)
4th	1.191:1 (25/21T)	1.136:1 (25/22T)	1.091:1 (24/22T)
5th	1.044:1 (24/23T)	0.958:1 (23/24T)	0.917:1 (22/24T)
6th	0.917:1 (22/24T)	0.808:1 (21/26T)	0.769:1 (20/26T)

Gear backlash – all pinions .... 0.044 – 0.140 mm (0.0017 – 0.0055 in)
Service limit .... 0.300 mm (0.0118 in)
Gearbox shaft OD – refer to Fig. 1.9 for details:
    Input shaft – at A .... 15.989 – 16.000 mm (0.6295 – 0.6299 in)
    Service limit .... 15.960 mm (0.6284 in)
    Input shaft – at B .... 19.959 – 19.980 mm (0.7858 – 0.7866 in)
    Service limit .... 19.920 mm (0.7843 in)
    Output shaft – at C .... 16.966 – 16.984 mm (0.6680 – 0.6687 in)
    Service limit .... 16.940 mm (0.6669 in)
    Output shaft at D .... 21.959 – 21.980 mm (0.8645 – 0.8654 in)
    Service limit .... 21.920 mm (0.8630 in)
    Output shaft – at E .... 19.974 – 19.987 mm (0.7864 – 0.7869 in)
    Service limit .... 19.940 mm (0.7850 in)
Gear pinion ID:
    Input shaft 5th and 6th gears .... 20.020 – 20.041 mm (0.7882 – 0.7890 in)
    Service limit .... 20.100 mm (0.7913 in)
    Output shaft 1st gear .... 19.020 – 19.041 mm (0.7488 – 0.7496 in)
    Service limit .... 19.100 mm (0.7520 in)
    Output shaft 3rd gear .... 24.020 – 24.041 mm (0.9457 – 0.9465 in)
    Service limit .... 24.100 mm (0.9488 in)
    Output shaft 2nd and 4th gears .... 22.020 – 22.041 mm (0.8669 – 0.8678 in)
    Service limit .... 22.100 mm (0.8701 in)
Output shaft gear pinion centre bush dimensions:
    OD – 1st gear .... 18.984 – 18.995 mm (0.7474 – 0.7478 in)
    Service limit .... 18.900 mm (0.7441 in)
    ID – 1st gear .... 17.016 – 17.034 mm (0.6699 – 0.6706 in)
    Service limit .... 17.100 mm (0.6732 in)
    OD – 2nd gear .... 21.984 – 22.005 mm (0.8655 – 0.8663 in)
    Service limit .... 21.900 mm (0.8622 in)
    ID – 2nd gear .... 20.020 – 20.041 mm (0.7882 – 0.7890 in)
    Service limit .... 20.100 mm (0.7913 in)
    OD – 3rd gear .... 23.984 – 23.993 mm (0.9443 – 0.9446 in)
    Service limit .... 23.900 mm (0.9409 in)
    ID – 3rd gear .... 22.020 – 22.041 mm (0.8669 – 0.8678 in)
    Service limit .... 22.100 mm (0.8701 in)
Selector drum OD – at left-hand end .... 11.966 – 11.984 mm (0.4711 – 0.4718 in)
Service limit .... 11.940 mm (0.4701 in)
Selector fork claw end thickness .... 4.93 – 5.00 mm (0.1941 – 0.1969 in)
Service limit .... 4.50 mm (0.1772 in)
Selector fork bore ID:
    Input shaft fork .... 10.000 – 10.018 mm (0.3937 – 0.3944 in)
    Service limit .... 10.050 mm (0.3957 in)
    Output shaft forks .... 12.000 – 12.018 mm (0.4724 – 0.4732 in)
    Service limit .... 12.050 mm (0.4744 in)
Selector fork shaft OD:
    For input shaft fork .... 9.972 – 9.987 mm (0.3926 – 0.3932 in)
    Service limit .... 9.950 mm (0.3917 in)
    For output shaft forks .... 11.966 – 11.984 mm (0.4711 – 0.4718 in)
    Service limit .... 11.940 mm (0.4701 in)

Final drive

	MBX125	MTX125	MTX200
Type	Chain and sprockets	Chain and sprockets	Chain and sprockets
Reduction ratio	3.000:1 (39/13T)	3.667:1 (55/15T)	3.231:1 (42/13T)
Chain size	520 (5/8 x 1/4)	428 (1/2 x 5/16)	520 (5/8 x 1/4)
Number of links	102	130	104

Drive chain – length of 40 links:
    MBX125, MTX200 .... 635 mm (25.0 in)

Service limit	648 mm (25.5 in)
MTX125	508 mm (20.0 in)
Service limit	518 mm (20.4 in)

Torque wrench settings

Component	kgf m	lbf ft
Cylinder head cover retaining nuts	0.8 – 1,.2	6 – 9
Cylinder head retaining nuts	2.0 – 2.4	14.5 – 17
Cylinder barrel retaining nuts	2.0 – 2.4	14.5 – 17
Inlet stub mounting bolts	0.8 – 1.2	6 – 9
Crankcase and crankcase cover bolts	0.8 – 1.2	6 – 9
Oil pump mounting bolt	0.8 – 1.2	6 – 9
ATAC valve assembly and operating spindle mountings – MTX200	1.8 – 2.0	13 – 14.5
Water pump cover and body bolts	0.8 – 1.2	6 – 9
Water pump impeller	1.3 – 1.7	9.5 – 12
Clutch centre retaining nut	6.0 – 7.0	43 – 50.5
Primary drive gear retaining nut	6.0 – 7.0	43 – 50.5
Selector drum detent roller arm pivot bolt	0.8 – 1.2	6 – 9
Alternator rotor retaining nut	6.0 – 7.0	43 – 50.5
Pulser coil mounting bolts	0.8 – 1.2	6 – 9
Bearing retainer bolts	0.8 – 1.2	6 – 9
Transmission oil drain plug	2.0 – 2.5	14.5 – 18
Kickstart ratchet guide plate mounting bolts	1.0 – 1.6	7 – 11.5
Engine top mounting bolts:		
MBX125	2.0 – 2.5	14.5 – 18
MTX200	2.4 – 3.0	17 – 22
Engine mounting bolt retaining nuts – MBX125	5.5 – 7.0	40 – 50.5
Engine front and upper rear mounting bolt retaining nuts – MTX125/200	4.0 – 5.0	29 – 36
Engine lower rear mounting bolt retaining nut – MTX 125/200	4.5 – 6.0	32.5 – 43
Exhaust pipe/cylinder barrel retaining nuts	1.8 – 2.5	13 – 18
Exhaust pipe rear mounting – MBX125	2.0 – 2.4	14.5 – 17
Exhaust pipe and silencer mounting bolts – MTX125/200	1.8 – 2.5	13 – 18
Crankcase backplate mounting bolts – MTX125/200	0.8 – 1.2	6 – 9
Radiator mounting bolts:		
MBX125	0.9 – 1.3	6.5 – 9.5
MTX125/200	0.8 – 1.2	6 – 9
Radiator cover/belly fairing mounting nuts – MBX125	0.8 – 1.2	6 – 9
Belly fairing mounting screws – MBX125	0.8 – 1.1	6 – 8
Coolant drain plugs	0.8 – 1.2	6 – 9
Gearchange pedal pinch bolt	0.8 – 1.2	6 – 9
Kickstart lever pinch bolt	2.0 – 2.4	14.5 – 17

1 General description

The crankshaft is a built-up assembly which rotates on two ball journal main bearings and had a connecting rod with needle roller big- and small-ends. A flywheel generator is fitted on its left-hand end, a gear on its right-hand end providing the drive for the clutch and oil pump, and on the MTX200 only, the ATAC mechanism. The clutch is a wet multi-plate type mounted on the right-hand end of the gearbox input shaft, the gearbox itself being of the constant mesh type and having six ratios.

To cancel out most of the vibration inherent in any single cylinder engine, a single shaft balancer is fitted, this being driven via an anti-backlash gear mounted on a separate drive shaft which is itself driven from the clutch outer drum. The same drive shaft passes through the gearbox to drive the water pump via a second pair of gears on early models; on MTX125 RW-H and L models one gear is deleted and the pump is driven off the end of the balancer shaft.

In spite of the many features (such as water-cooling and the anti-backlash gears) employed to reduce noise levels and to keep the unit as compact as possible, the engine is straightforward to work on and requires only a minimum of special tools.

2 Operations with the engine gearbox unit in the frame

The following items can be removed with the engine/gearbox unit in the frame:

a) Cylinder head, barrel and piston, carburettor and reed valve
b) Oil pump
c) Clutch assembly and primary drive gear pinion
d) ATAC mechanism – MTX200 only
e) Gear selector mechanism external components
f) Kickstart idler gears
g) Balancer/water pump drive idler shaft
h) Alternator, ignition pulser coil and neutral indicator switch
i) Gearbox sprocket
j) Water pump impeller and mechanical seal

3 Operations with the engine/gearbox unit removed from the frame

It is necessary to remove the engine/gearbox unit from the frame and to separate the crankcase halves to gain access to the following components:

a) Crankshaft, main bearings and oil seals
b) Gearbox clusters and bearings
c) Gear selector drum and forks
d) Balancer shaft
e) Kickstart mechanism
f) Water pump shaft, drive gears and bearings – except MTX125 RW-H and L

Note that while it is possible to remove and refit the crankshaft right-hand main bearing oil seal without separating the crankcase halves, this requires a great deal of care and is not recommended for the reasons given in Section 14.

4 Removing the engine/gearbox unit from the frame

1   If the machine is dirty, wash it thoroughly before starting any major dismantling work. This will make work much easier and will rule out the risk of caked-on lumps of dirt falling into some vital component.

2   Drain the transmission oil as described in Routine Maintenance.

While the oil is draining remove the seat, the side panels, the fuel tank, the crankcase backplate or belly fairing and the radiator shroud(s). Take care when removing the side panels and radiator shroud(s); they are hooked in place at certain points and must be disengaged carefully if the hooks are not to be broken.

3   Note that whenever any component is removed, all mounting nuts, bolts or screws should be refitted in their original locations with their respective washers and mounting rubbers and/or spacers.

4   Working as described in Chapter, 2, drain all the coolant from the cooling system. Tilt the machine to the left to ensure that as much as possible is removed.

5   Work is made much easier if the machine is lifted to a convenient height on a purpose-built ramp or platform constructed of planks and concrete blocks. Ensure that the wheels are chocked with wooden blocks so that the machine cannot move and that it is securely tied down so that it cannot fall, also that it is supported firmly on its stand.

6   Disconnect the battery (negative terminal first) to prevent any risk of short circuits. If the machine is to be out of service for some time, remove the battery and give it regular refresher charges as described in Chapter 7.

7   Remove the exhaust system. On MBX125 models, remove the single rear mounting bolt and the two front mounting nuts and withdraw the complete exhaust pipe. On MTX125/200 models remove its single mounting bolt and pull rearwards the silencer/tailpipe unit, then remove the single mounting bolt and two front mounting nuts and manoeuvre the exhaust pipe clear of the frame. Discard the exhaust port gasket.

8   Marking its shaft so that it can be refitted in the same position, remove fully the kickstart lever pinch bolt and pull the lever off the shaft splines. Remove its mounting screws and withdraw the oil pump cover. Disconnect the suppressor cap from the spark plug and the lead from the temperature gauge sender unit.

9   Disconnect the oil tank/pump feed pipe at the pump union or at the junction between the rubber and the clear plastic pipes, and plug the pipe with a screw or bolt of suitable size to prevent the loss of oil and the entry of air into the system. Release the feed pipe from any clamps or ties and tape it to the frame well clear of the engine/gearbox unit. Slacken fully the oil pump cable adjusting nut and locknut, release the adjuster from the crankcase cover and release the cable end nipple from the pump control lever. Unscrew the carburettor top, withdraw the throttle slide assembly and tape the throttle/oil pump cable assembly to the frame top tubes out of harm's way. Wrap the slide and needle in clean rag to prevent damage.

10   Removing fully its retaining screw, pull the tachometer cable out of its housing in the crankcase right-hand cover. Slacken fully its adjusting nut and locknut, release the clutch cable adjuster from its bracket and disengage the cable end nipple from the release lever. Allow the clutch and tachometer cables to hang down the frame front downtube well clear of the engine.

11   If the manufacturer's marks cannot be seen, mark its shaft so that it can be refitted in the same position, remove fully its pinch bolt and pull the gearchange pedal or linkage front arm (as appropriate) off the gearchange shaft splines. Remove its retaining screws and withdraw the crankcase left-hand cover.

12   Apply the rear brake hard to prevent rotation and slacken both gearbox sprocket retaining bolts. Remove them fully, rotate the locking plate until it can be pulled off the shaft splines, then pull off the gearbox sprocket. It may be necessary to slacken off the chain adjusters to permit this. Disengage the sprocket from the chain and allow the chain to hang over the swinging arm pivot.

13   Disconnecting the top and bottom radiator hoses as described in Chapter 2, remove the radiator from the machine so that the hoses remain affixed to the radiator. Starting from the crankcase, trace the generator wiring lead up the frame front downtube, releasing it from any clamps or ties securing it to the frame, then disconnect all wires at the connectors joining them to the main wiring loom.

14   Slacken fully the clamps securing the carburettor to the air filter hose and inlet stub, then carefully twist the carburettor clear of its mountings and remove it. Where applicable, pull the crankcase breather tube out of the air filter casing.

15   On MTX125/200 models only, it is necessary to remove the final drive chain tensioner assembly and the rear brake pedal assembly to gain access to the engine lower rear mounting bolt.

16   The engine/gearbox unit should now be retained only by its mounting bolts; check carefully that all components have been withdrawn which might hinder the removal of the unit. On MBX125 models it was found necessary to remove the spark plug and the cylinder head cover to provide sufficient clearance for the engine to be removed; check whether this is the case on the machine being worked on.

17   On MBX125 and MTX200 models remove the bolt securing the engine top mounting plate to the cylinder barrel; while it is not strictly necessary to remove the plate, work is a lot easier if it is out of the way. Remove its two mounting bolts and withdraw it. On all models, remove the three retaining nuts from the engine mounting bolts then tap out first the front bolt, then the upper rear, and finally the lower rear bolt. Note the single spacer on the front bolt (MBX125) and the two spacers on the upper rear bolt (MTX125/200). If any bolt is locked in place with corrosion or dirt, apply a liberal quantity of penetrating fluid, allow time for it to work and then release the bolt by rotating it with a spanner before tapping it out with a hammer and drift. Lift the engine/gearbox unit out of the frame, removing it to the left.

4.2a Belly fairing (MBX125) is retained by two nuts at the front ...

4.2b ... and by two screws at the rear

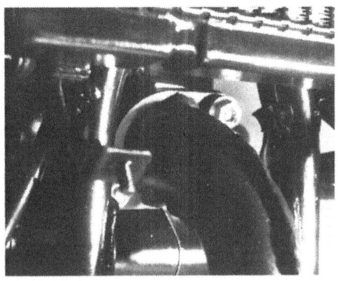

4.7a Remove two nuts to release exhaust pipe from cylinder barrel

4.7b MBX125 – exhaust is retained by a single bolt at the rear

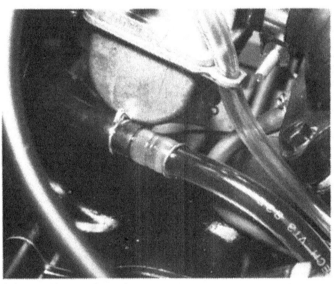

4.9a Oil tank/pump feed pipe can be disconnected at pump union or at junction shown

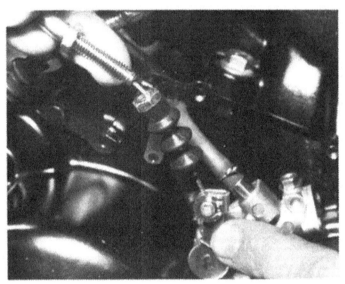

4.9b Disconnecting the oil pump control cable

4.10a Remove fully tachometer cable retaining screw ...

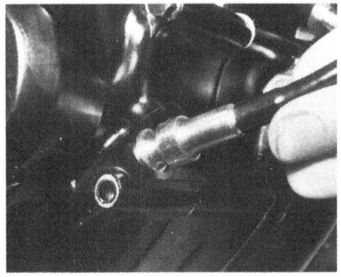

4.10b ... and pull cable out of its housing

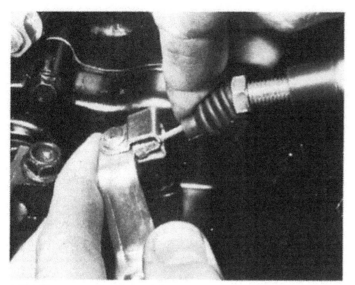

4.10c Disconnect the clutch cable

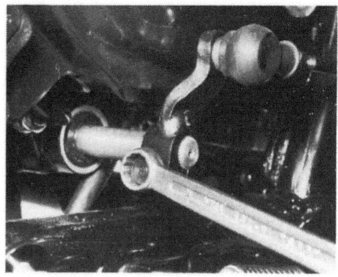

4.11 Mark lever and shaft to ensure correct refitting before removing gearchange lever

4.12 Remove retaining bolts and withdraw locking plate to release gearbox sprocket

4.15 MTX125/200 – remove chain tensioner assembly and brake pedal to reach lower rear engine mounting bolt

4.17a MBX125 and MTX200 – remove single bottom bolt to release engine top mounting plate – plate can be removed completely if necessary

4.17b Remove engine front mounting bolt – note single spacer (MBX125)

4.17c Remove engine rear upper ...

4.17d ... and rear lower mounting bolts to release engine unit

5   Dismantling the engine/gearbox unit: preliminaries

1   Before any dismantling work is undertaken, the external surfaces of the unit should be thoroughly cleaned and degreased. This will prevent the contamination of the engine internals, and will also make working a lot easier and cleaner. A high flash point solvent, such as paraffin (kerosene) can be used, or better still, a proprietary engine degreaser such as Gunk. Use old paintbrushes and toothbrushes to work the solvent into the various recesses of the engine castings. Take care to exclude solvent or water from the electrical components and inlet and exhaust ports. The use of petrol (gasoline) as a cleaning medium should be avoided, because the vapour is explosive and can be toxic if used in a confined space.
2   When clean and dry, arrange the unit on the workbench, leaving a suitable clear area for working. Gather a selection of small containers and plastic bags so that parts can be grouped together in an easily identifiable manner. Some paper and a pen should be on hand to permit notes to be made and labels attached where necessary. A supply of clean rag is also required.
3   Before commncing work, read through the appropriate section so that some idea of the necessary procedure can be gained. When removing the various engine components it should be noted that great force is seldom required, unless specified. In many cases, a component's reluctance to be removed is indicative of an incorrect approach or removal method. If in any doubt, re-check with the text.

6   Dismantling the engine/gearbox unit: removing the cylinder head, barrel and piston

1   These items can be removed with the engine in or out of the frame but in the former case the side panels, the seat, the radiator shrouds and (where fitted) the belly fairing, the fuel tank, the carburettor and the exhaust system must first be removed, the coolant must be drained and the suppressor cap, radiator hoses and temperature gauge sender unit wire must be disconnected. On MBX125 and MTX200 models the engine top mounting plates must be removed to allow the barrel to be withdrawn. Refer to Section 4 of this Chapter.
2   Disconnect the oil feed pipe from the inlet stub union and plug the pipe with a screw or bolt of suitable size to prevent the loss of oil and the entry of air or dirt into the system. Remove the four bolts securing the inlet stub to the rear of the barrel. Tapping very gently with a soft-faced mallet to break the seal of the two gaskets, lift away the inlet stub and reed valve assembly.

3   Unscrew the spark plug and remove the two nuts and washers which retain the cylinder head cover to the barrel. Lift away the cover, noting the presence of the two locating dowels and the two rubber seals; one around the spark plug boss, one sealing the cover/barrel joint.
4   Working in a diagonal sequence and by one turn at a time to avoid the risk of distortion, remove the five cylinder head retaining nuts and lift away the cylinder head. Discard the head gasket. If the head is stuck to the barrel do not attempt to lever it away. Break the joint by refitting the spark plug and turning the engine over, using its compression to release the head, or by tapping it gently with a soft-faced mallet. Be very careful when handling the cylinder head, it is a delicate casting and could easily be distorted or cracked by careless workmanship.
5   On MTX200 models, disconnect the ATAC valve. Displace its retaining clip and slide back the rubber cover. Unscrew the nut from the top of the operating spindle and withdraw the washer (where fitted), the collar, both parts of the connecting arm and the spring noting carefully which way round each is fitted, then remove the rubber cover. The valve assembly need not be removed from the cylinder unless required; refer to Chapter 3.
6   On all models, bring the piston to the top of its stroke, check that the water hose or pipe is removed, and remove the four cylinder barrel retaining nuts. Tap the barrel gently with a soft-faced mallet to break the seal and lift it just enough to expose the bottom of the piston skirt. Pack a wad of clean rag in the crankcase mouth to prevent dirt or debris from falling in, then lift the barrel away.
7   Use a sharp-pointed instrument or a pair of needle-nosed pliers to remove one of the gudgeon pin circlips, noting exactly how it is fitted, then press out the gudgeon pin far enough to clear the connecting rod and withdraw the piston. Push out the small-end bearing. Discard the used circlips and base gasket and obtain new ones for reassembly.
8   If the gudgeon pin is a tight fit in the piston, soak a rag in boiling water, wring it out and wrap it around the piston; the heat will expand the piston sufficiently to release its grip on the pin. If necessary, the pin may be tapped out using a hammer and drift, but take care to support firmly the piston and connecting rod while this is done.
9   The piston rings are removed by holding the piston in both hands and gently prising the ring ends apart with the thumb nails until the rings can be lifted out of their grooves and on to the piston lands, one side at a time. The rings can then be slipped off the piston and put to one side for cleaning and examination; do not forget the thin expander in the second piston ring groove. If the rings are stuck in their grooves by excessive carbon deposits use three strips of thin metal sheet to remove them, as shown in the accompanying illustration. Be careful as the rings are brittle and will break easily if overstressed.

53

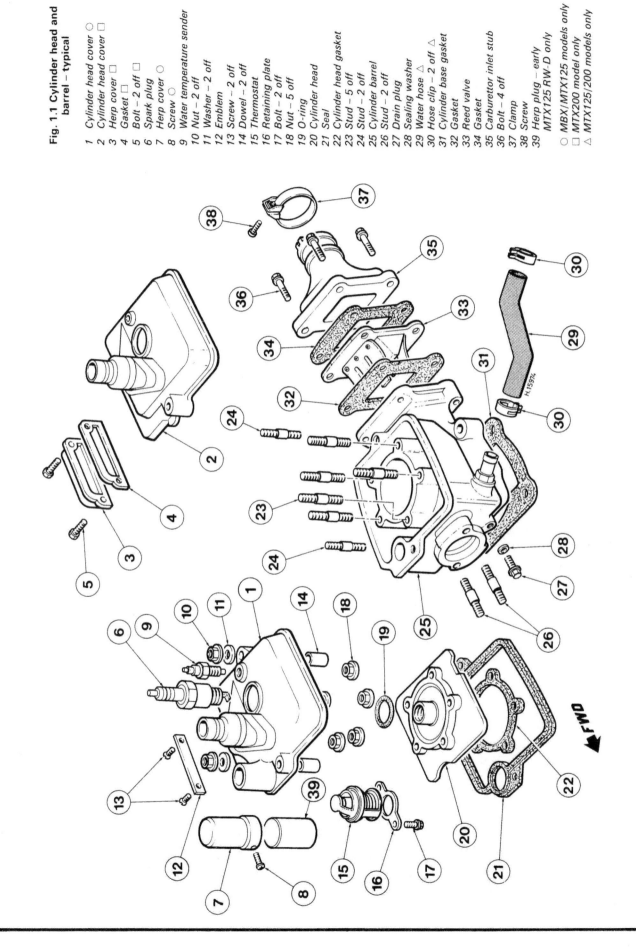

Fig. 1.1 Cylinder head and barrel – typical

1 Cylinder head cover ○
2 Cylinder head cover □
3 Herp cover □
4 Gasket □
5 Bolt – 2 off □
6 Spark plug
7 Herp cover ○
8 Screw ○
9 Water temperature sender
10 Nut – 2 off
11 Washer – 2 off
12 Emblem
13 Screw – 2 off
14 Dowel – 2 off
15 Thermostat
16 Retaining plate
17 Bolt – 2 off
18 Nut – 5 off
19 O-ring
20 Cylinder head
21 Seal
22 Cylinder head gasket
23 Stud – 5 off
24 Stud – 2 off
25 Cylinder barrel
26 Stud – 2 off
27 Drain plug
28 Sealing washer
29 Water hose △
30 Hose clip – 2 off △
31 Cylinder base gasket
32 Gasket
33 Reed valve
34 Gasket
35 Carburettor inlet stub
36 Bolt – 4 off
37 Clamp
38 Screw
39 Herp plug – early
   MTX125 RW-D only

○ MBX/MTX125 models only
□ MTX200 model only
△ MTX125/200 models only

6.2 Disconnect oil feed pipe and plug to prevent loss of oil

6.3 Remove two nuts and washers to release cylinder head cover

6.6a Water pump/cylinder barrel water hose must be disconnected before barrel can be withdrawn

6.6b Front two cylinder barrel retaining nuts are inside water jacket

6.6c Raise barrel as shown and pack crankcase mouth with clean rag

6.7a Use pliers to remove gudgeon pin circlips

6.7b Press out gudgeon pin until piston can be lifted away

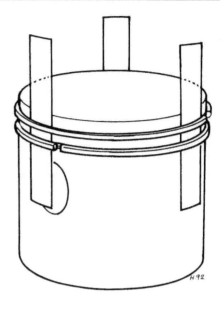

Fig. 1.2 Method of freeing gummed piston rings

7   Dismantling the engine/gearbox unit: removing the crankcase right-hand cover

1   This may be done with the engine in or out of the frame, but in the former case it will be necessary first to drain the oil, as described in Routine Maintenance and to disconnect the clutch, tachometer and oil pump cables, the oil tank/pump feed pipe, and to remove the kickstart lever from its shaft. Refer to Section 4 of this Chapter. Finally, disconnect the oil feed pipe from the inlet stub union and, on MTX200 models only, disconnect the ATAC spindle from the valve assembly, as described in the previous Section.

2   Remove its mounting bolt and carefully lift the oil pump assembly upwards out of the cover. This is necessary to prevent damage to its drive gears and to expose one cover retaining screw. For the same reasons, on MTX200 models only, unscrew its mounting nut and withdraw the ATAC operating spindle.

3   Working progressively and in a diagonal sequence from the outside in, remove the cover retaining screws. Store them in a cardboard template of the cover to provide a guide to their correct positions on reassembly; do not forget to store the clutch cable bracket with them.

4   Pull the cover away, tapping gently with a soft-faced mallet to break the seal. Do not attempt to lever the cover away; initial separation can be achieved easily by applying pressure, in its normal direction of operation, to the clutch release lever. Be careful that there are no thrust washers or other small components adhering to the cover. Peel off the gasket; unless the two locating dowels are firmly fixed in the crankcase they should be removed and stored with the cover.

7.2a Oil pump removal – unscrew single mounting bolt ...

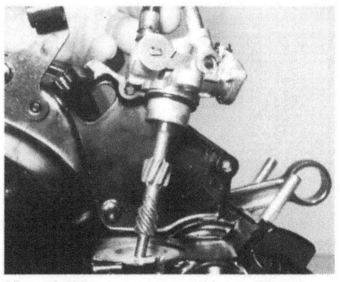

7.2b ... and withdraw complete pump assembly – note hidden cover retaining screw

8 Dismantling the engine/gearbox unit: removing the clutch and primary drive pinion

1   This can be done, whether the engine is in the frame or not, after the crankcase right-hand cover has been removed as described in the previous Section.

2   A special peg spanner will be required to slacken or tighten the clutch centre and primary drive pinion retaining nuts. If the Honda special tool, Part Number 07716-0010100, is not available, obtain a length of tubing with an approximate internal diameter of 20 mm (0.8 in) and a wall thickness of approximately 3 mm (0.12 in), then, referring to the accompanying illustration for details, cut away the segments shown to leave four protruding tangs which will engage with the slots in the nut. In the example shown in the accompanying photographs, this was then welded securely to an old socket so that a torque wrench could be used on reassembly.

3   The only alternative to this is to obtain a length of thick welding rod which can be bent into the shape of a flat 'U' and fitted across two of the opposing slots. A suitable-sized socket can then be tapped into place over the 'bridged' nut which can then be slackened or tightened in the usual way.

4   Before the nuts can be slackened, the crankshaft and clutch centre must be locked to prevent rotation. If the engine is in the frame select top gear and apply the rear brake hard, thus locking the clutch via the transmission; the crankshaft can be held by locking both parts of the clutch together as follows. With the lifter plate removed, refit two of

the clutch springs over their posts and tighten them down with two plain washers under the heads of the bolts instead of the lifter plate. Alternatively a holding tool can be applied to the alternator rotor to lock the crankshaft; if the two parts of the clutch are locked as just described (and shown in the accompanying photographs) the clutch centre nut can also be removed by this method. If the engine top half has been dismantled, lock the crankshaft by passing a close-fitting metal bar through the connecting rod small-end eye and rest it on two wooden blocks placed across the crankcase mouth.

5   All of the above methods are given as alternatives in case the manufacturer's special holding tools, Part Number 07923-KE10000 for the clutch and Part Number 07725-0030000 for the alternator rotor, are not available. To dismantle the clutch, proceed as follows.

6   Working by one turn at a time and in a diagonal sequence, remove the four clutch spring bolts and withdraw the clutch lifter plate, with its thrust bearing, and the clutch springs. Lock the clutch centre by one of the methods described above and unscrew the retaining nut. Withdraw the washer behind the nut. Holding two of the pressure plate spring posts, remove as a single assembly the clutch centre and the friction, plain and pressure plates.

7   Remove the first thrust washer, the clutch outer drum, its bearing sleeve and the second thrust washer from the input shaft.

8   To remove the primary drive gear, lock the crankshaft by one of the methods described above and unscrew the retaining nut. Remove the washer behind the nut and then pull off the primary drive gear, the ATAC shaft drive gear (MTX200 models only) and the spacer.

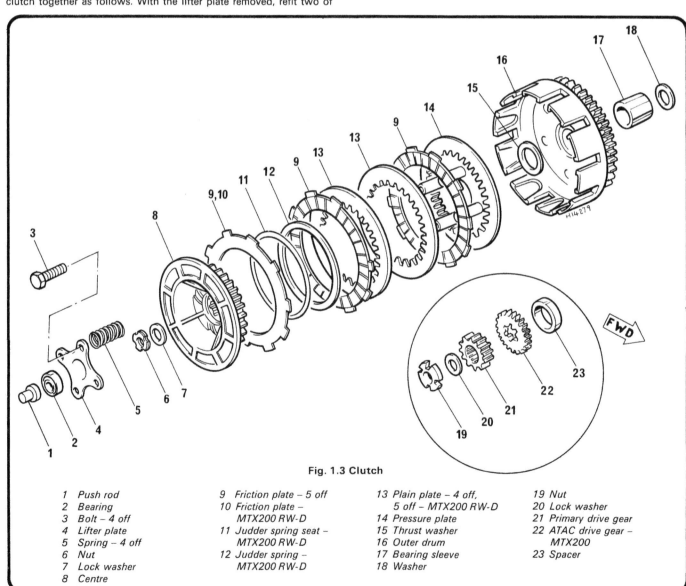

Fig. 1.3 Clutch

1 Push rod	9 Friction plate – 5 off	13 Plain plate – 4 off,	19 Nut
2 Bearing	10 Friction plate –	5 off – MTX200 RW-D	20 Lock washer
3 Bolt – 4 off	MTX200 RW-D	14 Pressure plate	21 Primary drive gear
4 Lifter plate	11 Judder spring seat –	15 Thrust washer	22 ATAC drive gear –
5 Spring – 4 off	MTX200 RW-D	16 Outer drum	MTX200
6 Nut	12 Judder spring –	17 Bearing sleeve	23 Spacer
7 Lock washer	MTX200 RW-D	18 Washer	
8 Centre			

8.2a Special peg spanner can be fabricated as described in text ...

8.2b ... and used to remove and refit clutch centre ...

8.2c ... and primary drive pinion retaining nuts – note methods employed to lock clutch centre and crankshaft

8.6a Slacken evenly its bolts and remove the clutch lifter plate

8.6b Removing the clutch plates

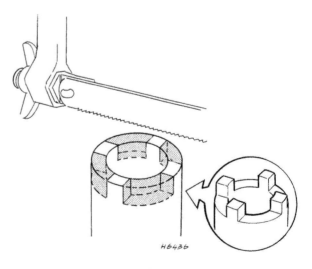

Fig. 1.4 Fabricated clutch peg spanner

9   Dismantling the engine/gearbox unit: removing the gear selector mechanism external components and the ancillary drive components

1   These can be removed whether the engine is in the frame or not, but only after the crankcase right-hand cover and clutch have been removed, as described in Sections 7 and 8, and after the gearchange pedal or linkage front arm has been withdrawn.

2   The balancer/water pump drive idler shaft can be pulled out of the crankcase after the clutch has been removed. Note the thrust washer behind the anti-backlash gears; a round pin or Woodruff key should be firmly fixed in the shaft left-hand end. Check that the pin or key is in place and do not lose it; if it has dropped into the crankcase the engine must be completely dismantled until it is found.

3   On MTX200 models only, the ATAC shaft can be pulled out of its crankcase bearing as soon as the crankcase right-hand cover has been removed. It may be necessary to rotate the clutch so that the shaft

driven gear teeth can be pulled past the teeth of the primary drive gear. Note the presence of any thrust washers (if fitted).

4   The kickstart idler gear and its bush can be pulled off the output shaft right-hand end once the clutch has been removed. If the kickstart pinion is to be removed, with the thrust washer, take care that the shaft is not disturbed; it is located in the crankcase only by the shoulder on the pinion left-hand surface.

5   To remove the external components of the gearchange mechanism, first remove the clutch, then remove from the selector drum right-hand end the circlip, the selector pin retaining plate and the selector detent cam, while pressing back the detent roller arm. Use a pair of pliers to pull out the five selector pins.

6   Remove its pivot bolt and withdraw the detent roller arm with its washer and return spring. Pull the claw arm assembly off the gearchange shaft end, noting the thrust washer behind it, then withdraw the thick spacer, the reduction gear and its pivot post. Pull out the gearchange shaft.

9.2 Do not lose Woodruff key on removing balancer/water pump drive idler shaft – note also the thrust washer

9.5 Press back detent roller arm to permit removal of selector cam and pin retaining plate

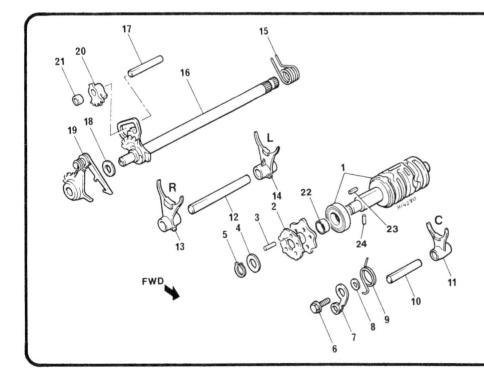

Fig. 1.5 Gear selector mechanism

1   Selector drum
2   Selector detent cam
3   Selector pin – 5 off
4   Selector pin retaining plate
5   Circlip
6   Bolt
7   Detent roller arm
8   Washer
9   Spring
10   Selector fork shaft
11   Selector fork C
12   Selector fork shaft
13   Selector fork R
14   Selector fork L
15   Return spring
16   Gearchange shaft
17   Pivot
18   Washer
19   Claw arm assembly
20   Reduction gear
21   Spacer
22   Spacer
23   Key – MTX200 RW-F
24   Key – all other models

10 Dismantling the engine/gearbox unit: removing the alternator

1   This can be done whether the engine is in the frame or not, but in the former case it will first be necessary to remove the belly fairing (MBX125) and the crankcase left-hand cover. If the stator is to be withdrawn it may be necessary to remove the radiator shroud and the radiator mounting bolts so that the wires can be disconnected from the main loom; it is not necessary to disconnect the hoses in such a case.
2   To lock the crankshaft to permit removal of the rotor retaining nut, select top gear and apply the rear brake hard, or apply a holding tool such as a strap wrench to the rotor itself. A more positive tool can be fabricated from three nuts and bolts and two strips of metal, as shown in the accompanying illustration. If the top end has been dismantled, as shown in the accompanying photograph, pass a close-fitting metal bar through the connecting rod small-end eye and support it on two wooden blocks placed across the crankcase mouth. With the crankshaft locked, remove the retaining nut and the washer behind it.
3   Remove the rotor using only a centre bolt flywheel puller, as shown

in the accompanying photograph. If this cannot be obtained from a Honda dealer under Part Number 07733-0010000, most good motorcycle dealers should be able to provide a cheaper pattern version which fits the majority of small-capacity Japanese machines. One of these tools should be considered essential; **no other method of rotor removal is recommended.**
4   Unscrew the tool centre bolt and screw the tool body anti-clockwise into the thread in the centre of the rotor (a **left-hand** thread is employed) until it seats firmly. Tighten the centre bolt down on to the crankshaft end and tap smartly on the bolt head with a hammer. The shock should jar the rotor free; if not, tighten the bolt further and tap again. Withdraw the rotor and its Woodruff key from the crankshaft.
5   Disconnect its lead from the neutral indicator switch and remove the three screws securing the alternator stator to the crankcase, then remove the two bolts securing the pulser coil to the crankcase. Note the way the wiring is routed and remove the single bolt securing the pulser coil wiring clamp. Withdraw the stator and pulser coil.
6   If desired, the neutral indicator switch can be unscrewed from the crankcase after the oil has been drained.

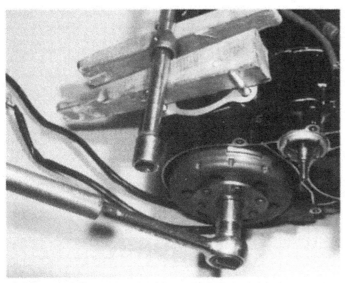

10.2 One possible method of locking crankshaft to permit removal of alternator rotor retaining nut

10.3 Remove rotor using only the type of tool shown – pattern version may be used if Honda service tool is not available

10.5a Remove three screws and two bolts (arrowed) to release alternator stator and pulser coil ...

10.5b ... noting that wiring clamp secured by single bolt (arrowed) must be removed ...

10.5c ... to permit removal of pulser coil

10.6 Neutral indicator switch may be unscrewed if required

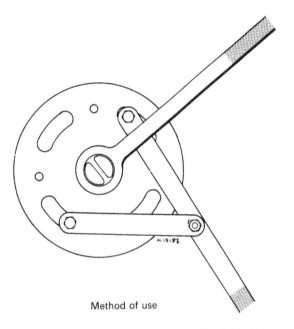

Method of use   Construction of tool

Fig. 1.6 Fabricated rotor holding tool

11 Dismantling the engine/gearbox unit: removing the water pump

1   These components can be removed with the engine in or out of the frame, but in the former case the belly fairing (MBX125), the radiator shroud(s) and the crankcase left-hand cover must be removed, the coolant must be drained and the hoses from the water pump to the radiator and to the cylinder barrel must be disconnected or their clamps slackened. Refer to Chapter 2.

2   Remove its three mounting bolts and withdraw the water pump cover taking care to mop up any spilt coolant. Note the sealing O-rings and the two locating dowels.

3   Either select top gear and apply the rear brake or apply a holding tool to the alternator rotor to lock the transmission while the impeller is unscrewed; if the transmission has been dismantled, on MBX125 and MTX125/200 RW-D, F models refit the balancer/water pump drive idler shaft temporarily and lock it by wedging a folded wad of rag or a wooden block between the teeth of the anti-backlash gear and the crankcase wall. On MTX125 RW-H and L models the balancer shaft

can be locked, if necessary, in a similar way. Unscrew the impeller **clockwise**. it employs a **left-hand** thread.

4   With the impeller removed, withdraw the copper washer. On MBX125 models ensure that the bolt clamping the pump/barrel water pipe to the pump body is removed, also the screw clamping the two pipes together. On MTX125 models ensure that the hose clamps are moved clear of the spigots on the pump body.

5   Remove the pump body mounting bolt that is situated inside the body, above the pump shaft; note that the body is also retained by another bolt at its rear end, but as this also retains the crankcase left-hand cover it should have been removed already by this stage. Manoeuvring it carefully to avoid damage to the hoses or pipes, withdraw the water pump body. The hoses or pipes (as appropriate) can be left attached to the radiator and cylinder barrel or removed with the pump body, whichever is most convenient.

6   Using a pair of pliers, pull out the water pump seal collar, noting the sealing O-ring around its outer circumference. Take care to extract the collar evenly or it may stick in its housing.

7   The water pump drive cannot be dismantled with the engine in the frame; the crankcase halves must be separated to reach it.

11.1 Crankcase left-hand cover top retaining screw also secures the rear of water pump body – remove crankcase cover before disturbing water pump

11.2 Remove three bolts (arrowed) to release water pump cover

11.3 Unscrew clockwise water pump impeller

11.4 Ensure coolant pipes and/or hoses are disconnected before removing water pump body

11.5 Note remaining water pump body retaining bolt

11.6 Water pump seal collar can be pulled from its housing – note sealing O-ring

12 Dismantling the engine/gearbox unit: separating the crankcase halves

1   The crankcases can be separated only after the engine/gearbox unit has been removed from the frame and all preliminary dismantling operations described in Section 4-11 of this Chapter have been carried out.
2   Make a final check that all components have been removed which might hinder crankcase separation. Make a cardboard template marking their respective positions and remove all crankcase fastening screws, slackening them progressively and in a diagonal sequence from the outside inwards.
3   Note that the crankcase left-hand half is to be lifted away leaving the crankshaft and gearbox components in the right-hand half.
4   Using only a soft-faced mallet, tap gently on the exposed ends of the crankshaft and gearbox shafts and all around the joint area of the two crankcase halves until initial separation is achieved. Lift off the half to be removed, ensuring it remains absolutely square so that the bearings do not stick on their respective shafts; tap gently on the shaft ends to assist removal. **Do not** use excessive force and **never** attempt to lever the cases apart. If undue difficulty is encountered, tap the cases back together and start again. If all else fails, take the assembly to a Honda Service Agent for the cases to be separated using special Honda tool Part Number 07931-KE10000.
5   If difficulty is encountered in achieving initial separation, it may be because corrosion has formed on the locating dowel pins. Apply a quantity of penetrating fluid to the joint area and inside the various mounting bolt passages, allow time for it to work and start again.
6   When the casing is removed, check that there are no loose components such as dowels or thrust washers which might drop clear and be lost. Any such components should be refitted in their correct locations; peel off and discard the gasket. Unless the two locating dowels are firmly fixed in the right-hand crankcase half they should be removed and stored in a safe place. Note that the water pump shaft (not fitted to MTX125 RW-H and L models) may stick to its bearing so that it is removed with the left-hand crankcase half.

13 Dismantling the engine/gearbox unit: removing the crankshaft and gearbox components

1   Except on MTX125 RW-H and L models, if the water pump shaft has stuck in its bearing in the crankcase left-hand half, it is removed as follows. Withdraw the retaining circlip, followed by the thrust washer and the pump driven gear itself. Remove the drive pin from the shaft which can then be tapped out to the left. Honda specify this procedure as the shaft and its bearing are listed as a single assembly; if this were the case the shaft/bearing could only be removed this way. In practice the shaft was found to be a slack enough fit on the bearing inner race to be pulled out by hand to the right. Either method may be possible on the machine being worked on.
2   Working on the right-hand crankcase half, first remove the kickstart assembly. Use a pair of pliers to unhook the return spring from its stop in the crankcase, then remove the thrust washer and pinion from the shaft right-hand end. Lift the assembly out to the left and put it to one side.
3   Using gently a soft-faced mallet, tap the balancer shaft out to the left.
4   To remove the gearbox components, pull out the shaft holding the single selector fork in front of the selector drum and withdraw the fork; if the fork has no identifying markings which can be used, mark its upper (ie left-hand) face with a spirit-based felt marker so that it can be readily identified and refitted the correct way round. Similarly withdraw their shaft and remove the two output shaft forks, then withdraw the selector drum. Tapping carefully with a soft-faced mallet on their right-hand ends, drive out both gearbox shafts as a single unit. As soon as they are free, tighten a rubber band over each shaft end to prevent the loss of any components, then check that all are correctly in place.
5   To remove the crankshaft, first attempt to lift it out of the crankcase half, using only a soft-faced mallet, tap **gently** on the shaft right-hand end to assist this, but take great care not to damage the oil pump drive worm. If this fails a slightly different approach will be necessary; **do not use excessive force**.
6   Firmly support the right-hand crankcase half on two wooden blocks so that the crankshaft hangs down but its end is clear of the work surface. Refit the primary drive gear retaining nut and protect the pump drive worm by placing a tubular drift (a socket spanner would be ideal) over it so that when the drift is tapped gently with a hammer the shock is transmitted to the nut rather than to the worm. **Do not use excessive force**; there is a high risk of damaging the nut or the crankshaft thread. Tap gently on the drift until the crankshaft is released; do not allow it to fall away, but hold it while the nut is removed and then lift it clear.
7   When using either of the above methods if the crankshaft cannot be removed easily with the minimum use of force, take the assembly to a Honda Service Agent for the crankshaft to be removed using the correct service tools.

12.3 Lift away left-hand crankcase half, leaving all components in right-hand half

12.6 Water pump shaft may stick in its bearing (so that it is withdrawn with the left-hand crankcase half) or it may drop clear, as shown – except MTX125 RW-H and L models

Fig. 1.7 Crankshaft assembly

1 Piston rings
2 Piston
3 Gudgeon pin
4 Small-end bearing
5 Circlip – 2 off
6 Crankshaft assembly
7 Right-hand crankshaft
8 Left-hand crankshaft
9 Left-hand main bearing
10 Oil pump drive gear
11 Right-hand main bearing
12 Connecting rod
13 Crank pin
14 Big-end bearing
15 Thrust washer – 2 off
16 Woodruff key
17 Balancer shaft – MBX125
   and MTX125/200 RW-D,
   RW-F
18 Balancer idler shaft
   assembly – MTX125 RW-F,
   H, L and MTX200 RW-F
19 Balancer idler shaft
20 Thrust washer
21 Wave washer
22 Sub gear A
23 Spring – 8 off
24 Balancer idler gear –
   MTX125/200 RW-F
25 Roller
26 Clip
27 Balancer idler gear –
   MTX125 RW-H, L
28 Damper – MTX125 RW-H, L
29 Balancer idler shaft
30 Washer
31 Sub gear spring
32 Sub gear B
33 Balancer idler gear
34 Sub gear A
35 Balancer idler shaft
   assembly – MBX125 and
   MTX125/200 RW-D
36 Balancer shaft –
   MTX125 RW-H, L
37 Bearing – MTX125 RW-H, L

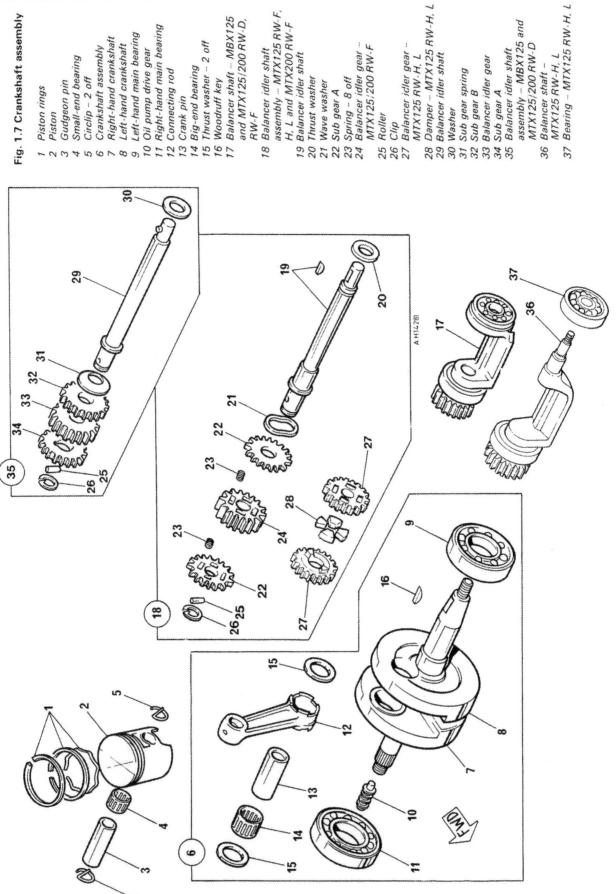

13.2 Use strong pliers to release kickstart return spring tension – do not risk damage or injury by allowing spring to fly back

14 Dismantling the engine/gearbox unit: removing oil seals and bearings

1   Before removing any oil seal or bearing, check that it is not secured by a retaining plate. If this is the case, use an impact driver or spanner, as appropriate, to release the securing screws or bolts and lift away the retaining plate.
2   Oil seals are easily damaged when disturbed and, thus, should be renewed as a matter of course during overhaul. Prise them out of position using the flat of a screwdriver and taking care not to damage the alloy seal housings; take care to note which way round the seals are fitted.
3   It is possible to remove the right-hand main bearing oil seal without separating the crankcases; either by screwing in two self-tapping screws so that the seal can be pulled out with two pairs of pliers, or by simply digging the seal out with a sharply-pointed instrument. This method is **not recommended** as it is difficult to carry out without scratching the seal housing or crankshaft or without damaging the main bearing. Furthermore, it is almost impossible to fit a new seal without damaging it and nothing can be done to trace and rectify the fault which caused the seal to fail in the first place.
4   The crankshaft and gearbox bearings are a press fit in their respective crankcase locations. To remove a bearing, the crankcase casting must be heated so that it expands and releases its grip on the bearing, which can then be drifted or pulled out.
5   To prevent casting distortion, it must be heated evenly to a temperature of about 100°C by placing it in an oven; if an oven is not available, place the casting in a suitable container and carefully pour boiling water over it until it is submerged.
6   Taking care to prevent personal injury when handling heated components, lay the casting on a clean surface and tap out the bearing using a hammer and a suitable drift. If the bearing is to be re-used apply the drift only to the bearing outer race, where this is accessible, to avoid damaging the bearing. In some cases it will be necessary to apply pressure to the bearing inner race; in such cases closely inspect the bearing for signs of damage before using it again. When drifting a bearing from its housing it must be kept square to the housing to prevent tying in the housing with the resulting risk of damage. Where possible, use a tubular drift such as a socket spanner which bears only on the bearing outer race; if this is not possible, tap evenly around the outer race to achieve the same result.
7   In some cases, particularly the water pump drive gear assembly, bearings are pressed into blind holes in the castings. These bearings

must be removed by heating the casting and tapping it face downwards on to a clean wooden surface to dislodge the bearing under its own weight. If this is not successful the casting should be taken to a motorcycle service engineer who has the correct internally expanding bearing puller. If a bearing sticks to its shaft on removal, it can only be safely removed using a knife-edged bearing puller. Note that in some cases, the balance shaft for example, the bearings are not listed as separate parts by Honda and are only available as part of the complete assembly. Unless there is an alternative source of replacement bearings of the correct size and type, there is little point in attempting to remove the bearings. This point should be borne in mind at all times.

15 Examination and renovation: general

1   Before examining the parts of the dismantled engine unit for wear it is essential that they should be cleaned thoroughly. Use a petrol/paraffin mix or a high flash-point solvent to remove all traces of old oil and sludge which may have accumulated within the engine. Where petrol is included in the cleaning agent normal fire precautions should be taken and cleaning should be carried out in a well ventilated place.
2   Examine the crankcase castings for cracks or other signs of damage. If a crack is discovered it will require a specialist repair.
3   Examine carefully each part to determine the extent of wear, checking with the tolerance figures listed in the Specifications section of this Chapter or in the main text. If there is any doubt about the condition of a particular component, play safe and renew.
4   Use a clean lint-free rag for cleaning and drying the various components. This will obviate the risk of small particles obstructing the internal oilways, and causing the lubrication system to fail.
5   Various instruments for measuring wear are required, including a vernier gauge or external micrometer and a set of standard feeler gauges. Both an internal and external micrometer will be required to check wear limits. Additionally, although not absolutely necessary, a dial gauge and mounting bracket is invaluable for accurate measurement of end float, and play between components of very low diameter bores – where a micrometer cannot reach.

After some experience has been gained the state of wear of many components can be determined visually or by feel and thus a decision on their suitability for continued service can be made without resorting to direct measurement.

14.1 Check first that seal or bearing is not located by retaining plate – plates must be removed to release bearing

14.2a Removing oil seals – take care not to damage seal housing ...

14.2b ... and note which way round seals are fitted before disturbing them

14.7 Some components, eg water pump drive gear/crankcase breather assembly shown, are fitted into blind holes – special equipment may be needed for removal

16 Examination and renovation: engine cases and covers

1   Small cracks or holes in aluminium castings may be repaired with an epoxy resin adhesive, such as Araldite, as a temporary measure. Permanent repairs can only be effected by argon-arc welding, and only a specialist in this process is in a position to advise on the economics or practicability of such a repair.
2   Damaged threads can be economically reclaimed by using a diamond section wire insert, of the Helicoil type, which is easily fitted after drilling and re-tapping the affected thread. Most motorcycle dealers and small engineering firms offer a service of this kind.
3   Sheared studs or screws can usually be removed with screw extractors, which consist of tapered, left-hand thread screws, of very hard steel. These are inserted by screwing anti-clockwise into a pre-drilled hole in the stud or screw, and usually succeed in dislodging the most stubborn stud or screw. If a problem arises which seems to be beyond your scope, it is worth consulting a professional engineering firm before condemning an otherwise sound casing. Many of these firms advertise regularly in the motorcycle papers.

17 Examination and renovation: bearings and oil seals

1   The crankshaft and gearbox bearings can be examined while they are still in place in the crankcase castings. Wash them thoroughly to remove all traces of oil, then feel for free play by attempting to move the inner race up and down, then from side-to-side. Examine the bearing balls or rollers and the bearing tracks for pitting or other signs of wear, then spin the bearing hard. Any roughness caused by defects in the bearing balls or rollers or in the bearing tracks will be felt and heard immediately.
2   If any signs of free play are discovered, or if the bearing is not free and smooth in rotation but runs roughly and slows down jerkily, it must be renewed. Bearing removal is described in Section 14 of this Chapter, and refitting in Section 28.
3   To prevent oil leaks occurring in the future, all oil seals and O-rings should be renewed whenever they are disturbed during the course of an overhaul, regardless of their apparent condition. This is particularly true of the main bearing oil seals, which are a weak point on any two-stroke.

18 Examination and renovation: cylinder head

1   Check that no cracks are evident, especially in the vicinity of the spark plug or stud holes, and remove all traces of carbon from the combustion chamber.
2   Check the condition of the thread in the spark plug hole. If it is damaged an effective repair can be made using a Helicoil thread insert. This service is available from most motorcycle dealers. Always use the correct plug and do not overtighten, see Routine Maintenance.
3   Leakage between the head and barrel will indicate distortion. Check the head by placing a straight-edge across several places on its mating surface and attempting to insert a 0.1 mm (0.004 in) feeler gauge between the two.
4   Remove excessive distortion by rubbing the head mating surface in a slow circular motion against emery paper placed on plate glass. Start with 200 grade paper and finish with 400 grade and oil. Do not remove an excessive amount of metal. If in doubt consult a Honda agent.
5   Note that most cases of cylinder head distortion can be traced to unequal tensioning of the cylinder head securing nuts or to tightening them in the incorrect sequence.

19 Examination and renovation: cylinder barrel

1 The usual indication of a badly worn cylinder barrel and piston is piston slap, a metallic rattle that occurs when there is little or no load on the engine.

2 Carefully remove the ring of carbon from the bore mouth so that bore wear can be accurately assessed and check the barrel/cylinder head mating surface as described in the previous Section. Clean all carbon from the exhaust port and all traces of old gasket from the cylinder base.

3 Examine the bore for scoring or other damage, particularly if broken rings are found. Damage will necessitate reboring and a new piston regardless of the amount of wear. A satisfactory seal cannot be obtained if the bore is not perfectly finished.

4 There will probably be a lip at the uppermost end of the cylinder bore which marks the limit of travel of the top of the piston ring. The depth of the lip will give some indication of the amount of bore wear that has taken place even though the amount of wear is not evenly distributed.

5 The most accurate method of measuring bore wear is by the use of a cylinder bore DTI (Dial Test Indicator) or a bore micrometer. Measure at the top (just below the wear ridge), middle and bottom of the bore, in line with the gudgeon pin axis and at 90° to it, avoiding port areas. Take six measurements in all noting each one carefully. Take the maximum figure obtained to indicate bore wear; if this is equivalent to, or greater than the service limit given in the Specifications Section of this Chapter, the barrel must be rebored and the next oversize piston and rings must be obtained. This can be confirmed by subtracting the overall diameter of the piston from the maximum cylinder bore figure obtained. If the difference calculated exceeds 0.16 mm (0.006 in) the bore can be regarded as excessively worn, assuming that the piston is unworn (see Section 20).

6 Alternatively, insert the piston into the bore in its correct position with the skirt just below the wear ridge, ie at the point of maximum wear. Measure the gap using feeler gauges. If the piston/cylinder clearance is 0.16 mm (0.006 in) or more the barrel must be rebored. It must be stressed that this can only be a guide to the degree of bore wear alone unless the piston is known to be unworn; have the barrel measured accurately by a competent Honda Service Agent or similar expert before a rebore is undertaken.

7 After a rebore has been carried out, the reborer should hone the bore lightly to provide a fine cross-hatched surface so that the new piston and rings can bed in correctly. Also the edges of the ports should be chamfered first with a scraper, then with fine emery, to prevent the rings from catching on them and breaking.

8 If a new piston and/or rings are to be run in a part-worn bore the surface should be prepared first by glaze-busting. This involves the use of a special honing attachment with (usually) an electric drill to provide a surface similar to that described above. Most motorcycle dealers have such equipment and will be able to carry out the necessary work for a small charge.

20 Examination and renovation: piston and piston rings

1 Disregard the existing piston and rings if a rebore is necessary; they will be replaced with oversize items. Note also that it is considered a worthwhile expense by many mechanics to renew the piston rings as a matter of course, regardless of their apparent condition.

2 Measure the piston diameter at right angles to the gudgeon pin axis, at a point 10 mm (0.4 in) above the base of the skirt. If the piston is worn to the service limit given in the Specifications Section of this Chapter or beyond, it must be renewed. Note that oversize pistons and rings are available in two sizes: 0.25 mm (0.010 in) and 0.50 mm (0.020 in). The amount of oversize is stamped on the piston crown; do not forget to include it in any calculations.

3 Note that the manufacturer's alternative method of determining whether a rebore is necessary is to subtract the piston overall diameter from the maximum bore internal diameter measurement obtained. Since it is possible for the piston to have worn at a greater rate than the bore, this can only be accurate if the piston measurement is that of an unworn component (ie new). If this method is used, always have your findings checked by an expert before undertaking a rebore; it may suffice to renew the piston alone.

4 Piston wear usually occurs at the skirt, especially on the forward face, and takes the form of vertical score marks. Reject any piston which is badly scored or which has been blackened as the result of the blow-by of gas. Slight scoring of the piston can be removed by careful use of a fine swiss file. Use chalk to prevent clogging of the file teeth and the subsequent risk of scoring. If the ring locating pegs are loose or worn, renew the piston.

5 The gudgeon pin should be a firm press fit in the piston. Check for scoring on the bearing surfaces of each part and where damage or wear is found, renew the part affected. If the equipment is available the components can be measured for wear; any item found to be worn to the service limit specified should be renewed. The circlip retaining grooves must be undamaged; renew the piston rather than risk damage to the bore through a circlip becoming detached. Discard the circlips themselves; these should **never** be reused.

6 Any build-up of carbon in the ring grooves can be removed by using a section of broken piston ring, the end of which has been ground to a chisel edge. Using a feeler gauge, measure each ring to groove clearance, where applicable. Renew the piston if the measurement obtained exceeds that given in Specifications and if the rings themselves appear unworn.

7 Measure ring wear by inserting each ring into part of the bore which in unworn and measuring the gap between the ring ends with a feeler gauge. If the measurement exceeds that given in Specifications, renew the ring. Use the piston crown to locate the ring squarely in the bore approximately 20 mm (0.8 in) from the gasket surface.

8 Reject any rings which show discoloured patches on their mating surfaces; these should be brightly polished from firm contact with the cylinder bore.

9 Do not assume when fitting new rings that their end gaps will be correct; the installed end gap must be measured as described above to ensure that it is within the specified tolerances; if the gap is too wide another piston ring set must be obtained (having checked again that the bore is within specified wear limits), but if the gap is too narrow it must be widened by the careful use of a fine file.

21 Examination and renovation: crankshaft assembly

1 Big-end failure is characterised by a pronounced knock which will be most noticeable when the engine is worked hard. The usual causes of failure are normal wear, or failure of the lubrication supply. In the case of the latter, the noise will become apparent very suddenly, and will rapidly worsen.

2 Check for wear with the crankshaft set in the TDC (top dead centre) position, by pushing and pulling the connecting rod. No discernible movement will be evident in an unworn bearing, but care must be taken not to confuse end float, which is normal, and bearing wear.

3 If a dial gauge is available, set its pointer against the big end eye. Measure the amount of radial play of the connecting rod. Renew the big-end bearing if the measurement exceeds 0.05 mm (0.002 in).

4 Push the connecting rod to one side and use a feeler gauge to measure the big-end clearance. Renew the big-end bearing, the crank pin and its washers and the connecting rod if the clearance exceeds that specified.

5 Set the crankshaft on V-blocks positioned on a completely flat surface and measure the amount of deflection with the dial gauge pointer set against the inboard end of either mainshaft. Renew the crankshaft if the measurement exceeds 0.10 mm (0.004 in).

6 Push the small-end bearing into the connecting rod eye and push the gudgeon pin through the bearing. Hold the rod steady and feel for movement between it and the pin. If movement is felt, renew the pin, bearing or connecting rod as necessary so no movement exists. If the equipment is available, the gudgeon pin and small-end eye can be measured to assess which one is worn. Renew the bearing if its roller cage is cracked or worn.

7 Do not attempt to dismantle the crankshaft assembly, this is a specialist task. If a fault is found, return the assembly to a Honda agent who will supply a new or service-exchange item.

8 The oil pump drive worm cannot be extracted easily from the crankshaft end and should be left in place unless absolutely necessary. Although it is listed separately it also is sold as part of the complete crankshaft, should a new assembly be purchased.

20.2 Measuring piston diameter

20.7 Measuring piston ring installed end gap

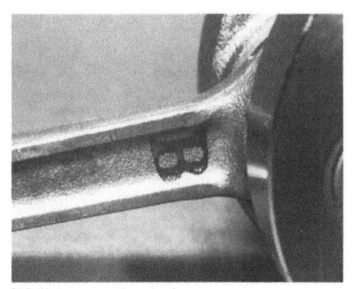

21.7a Marks on connecting rod and ...

21.7b ... crankpin indicate that big-end bearing components are carefully matched on assembly – repair is a task for the expert alone

22 Examination and renovation: primary drive, balancer/water pump drive and balancer shaft

1 The primary drive consists of a crankshaft pinion which engages a large gear mounted on the inner face of the clutch drum. Both components are relatively lightly loaded and will not normally wear until a very high mileage has been covered.

2 If wear or damage is discovered it will be necessary to renew the component concerned. In the case of the large driven gear it will be necessary to purchase a complete clutch drum because the two items form an integral unit and cannot be obtained separately. Note that the large driven gear/clutch outer drum assembly has a shock absorber fitted. Check for play in this by holding the clutch outer drum and attempting to twist or rotate the primary driven gear backwards and forwards. Unfortunately no figures are given with which to assess the state of the shock absorber unit, and it will therefore be a matter of experience to decide whether renewal is necessary or not. Seek an

expert opinion if any doubt exists about the amount of play discovered.

3 Check that there is no obvious damage to the balancer shaft such as chipped or broken teeth or worn bearings; if any wear or damage is found the shaft must be renewed as a complete assembly.

4 Measure the balancer/water pump drive idler shaft at its bearing surfaces and renew it if it has worn to the service limits specified. Check that the shaft is straight and otherwise undamaged. The anti-backlash gear consists of two narrow pinions spring-loaded against the central, thicker pinion; the two narrow pinions are set to run slightly out of alignment with the main pinion, thus taking up any backlash (and eliminating noise) in the gear train.

5 Check that the gears are not chipped, damaged or worn and that the narrow pinions are stiff to turn against spring pressure; do not dismantle the assembly unless absolutely necessary. If dismantling is necessary some means must be devised of clamping the right-hand narrow pinion against the shaft left-hand end so that the retaining clip can be removed. The pressure can then be relaxed slowly so that the assembly can be dismantled.

22.5 Do not attempt to dismantle anti-backlash gears unless absolutely necessary

23 Examination and renovation: clutch

1   Clean the clutch components in a petrol/paraffin mix whilst observing the necessary fire precautions.
2   Overworn friction plates will cause clutch slip. Measure the thickness of each plate and renew it if the thickness is less than the limit given in Specifications. On MTX200 RW-D models, renew the judder spring and seat if either is damaged or worn.
3   Check the friction plate tangs and the drum slots for indentations caused by clutch chatter. If slight, the damage can be removed with a fine file, otherwise renewal if necessary.
4   Check all plates for distortion. Lay each one on a sheet of plate glass and attempt to insert a feeler gauge beneath it. Refer to Specifications for maximum allowable warpage.

5   Inspect each plain plate for scoring and signs of overheating in the form of blueing. Remove slight damage to the plate tangs and hub slots with a fine file, otherwise renewal is necessary.
6   Examine the pressure plate for cracks and overheating. Look for hairline cracks around the base of each spring location and around the centre boss. Check for excessive distortion.
7   Measure the free length of each spring and renew all of them if one has set to less than the limit given in Specifications.
8   Measure the clutch outer drum centre bush, the input shaft and the bearing sleeve, renewing any component that is found to be excessively worn (see Specifications).
9   Check the components of the release mechanism for wear or damage and renew as necessary. Grease thoroughly on reassembly.

24 Examination and renovation: kickstart assembly

1   Check the condition of the kickstart components. If slipping has been encountered, a worn ratchet and pinion will invariably be traced as the cause. Any other damage or wear to the components will be self-evident. If either the ratchet or pinion is found to be faulty, the components must be replaced as a pair. Examine the kickstart return spring, which should be renewed if there is any doubt about its condition.
2   To dismantle the assembly, first slide out the return spring guide, then disengage the spring inner end from the hole in the shaft and remove the spring. This is followed by a large thrust washer, a light coil spring and the kickstart ratchet.
3   If the equipment is available, measure the pinion and idler gear internal diameters, the outside diameters of their shafts and the dimensions of the idler gear centre bush; renew any component that is found to be excessively worn.
4   The components are refitted by reversing the dismantling sequence. Note that when refitting the ratchet to the shaft, the punch mark on each should be in alignment with the other. If this alignment is not correct, the kickstart will not operate correctly.
5   In general, kickstart mechanisms are reliable devices and do not wear quickly unless a malfunction occurs, or engine oil level is allowed to fall. Look for chipped or broken teeth and excessive wear, replacing the ratchet and the kickstart pinion if necessary.

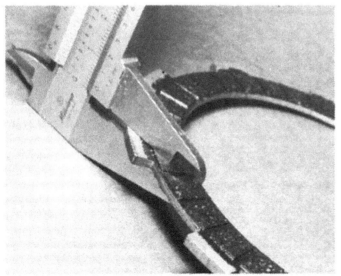

23.2 Measuring clutch friction plate thickness

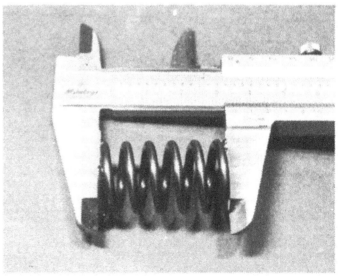

23.7 Measuring clutch spring free length

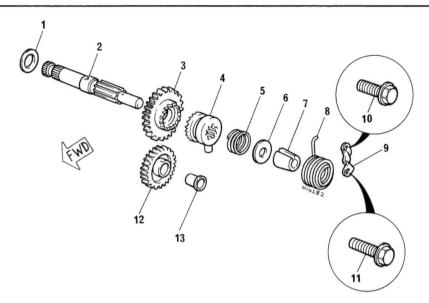

Fig. 1.8 Kickstart mechanism

1   Thrust washer
2   Spindle
3   Pinion
4   Ratchet
5   Spring
6   Thrust washer
7   Spring guide
8   Return spring
9   Ratchet guide plate
10 Bolt
11 Bolt
12 Idler gear
13 Bush

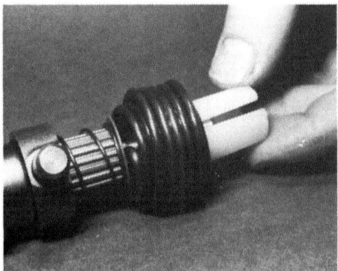

24.2a Dismantling the kickstart assembly – withdraw the spring guide and return spring ...

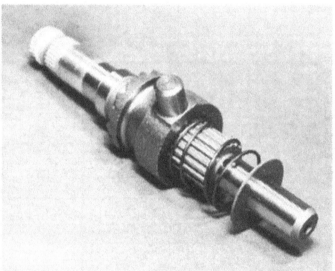

24.2b ... then remove the thrust washer, coil spring and kickstart ratchet

24.4 On reassembly, fit ratchet onto shaft so that punch mark on ratchet aligns with mark on shaft – if shaft mark is not visible, align ratchet mark with return spring hole

25 Examination and renovation: gearbox components

1   Give the gearbox components a close visual inspection for signs of wear or damage such as broken or chipped teeth, worn dogs, damaged or worn splines and bent selectors. Replace any parts found unserviceable because they cannot be reclaimed in a satisfactory manner.

2   The gearbox shafts are unlikely to sustain damage unless the lubricating oil has been run low or the engine has seized and placed an unusually high loading on the gearbox. Check the surfaces of the shafts, especially where a pinion turns on them, and renew the shafts if they are scored or have picked up. Referring to the accompanying illustration, measure the shafts at the points indicated and renew them if they are worn to the service limits specified. The procedure for dismantling and rebuilding the gearbox shaft assemblies is given in the following Section.

3   Examine the gear selector claw assembly noting that worn or rounded ends on the claw can lead to imprecise gear selection. The springs in the selector mechanism and the detent or stopper arm should be unbroken and not distorted or bent in any way.

4   Examine the selector forks carefully, ensuring that there is no sign of scoring on the bearing surface of either of their claw ends, their bores or their selector drum guide pins. Renew the forks if their claw

ends are worn to the service limits specified. Check for any signs of cracking around the edges of the bores or at the base of the fork arms.

5   Check each selector fork shaft for straightness by rolling it on a sheet of plate glass and checking for any clearance between the shaft and the glass with feeler gauges. A bent shaft will cause difficulty in selecting gears. There should be no sign of any scoring on the bearing surface of the shaft or any discernible play between each shaft and its selector fork(s). This can be checked by careful measurement of the components and the comparison of the results obtained with the service limits specified.

6   The tracks in the selector drum should not show signs of undue wear or damage. Check also by measurement, where applicable, that the selector drum bearing surfaces are unworn; renew the drum if it is worn or damaged.

7   Note that certain pinions have a bush fitted within their centres. If any one of these bushes appears to be over-worn or in any way damaged, its dimensions can be compared with those given in the Specifications Section of this Chapter. Measure also the internal diameter of the gear pinions and renew any component found to be excessively worn.

8   Finally, carefully inspect the splines of both shafts and pinions for any signs of wear, hairline cracks or breaking down of the hardened surface finish. If any one of these defects is apparent, then the offending component must be renewed. It should be noted that damage and wear rarely occur in a gearbox which has been properly used and correctly lubricated, unless a very high mileage has been covered. Pinion tooth wear can only be assessed using a dial gauge and stand to measure the tooth backlash. If the equipment is available,

and the shafts can be mounted securely so that the backlash can be measured accurately; any pinion or pair of pinions with backlash of 0.3 mm (0.012 in) or more must be renewed.

9   Clean the gearbox sprocket thoroughly and examine it closely, paying particular attention to the condition of the teeth. The sprocket should be renewed if the teeth are hooked, chipped, broken or badly worn. It is considered bad practice to renew one sprocket on its own; both drive sprockets should be renewed as a pair, preferably with a new final drive chain. Examine the splined centre of the sprocket for signs of wear. If any wear is found, renew the sprocket as slight wear between the sprocket and shaft will rapidly increase due to the torsional forces involved. Remember that as the output shaft will probably wear in unison with the sprocket, it is therefore necessary to carry out a close inspection of the shaft splines.

10 Carefully examine the gear selector components. Any obvious signs of damage, such as cracks, will mean that the part concerned must be renewed. Check that the springs are not weak or damaged and examine all points of contact, eg gearchange shaft teeth and selector claw arm, for signs of excessive wear. If doubt arises about the condition of any component it must be compared with a new part to assess the amount of wear that has taken place, and renewed if found to be damaged or excessively worn. Do not forget to check the pins set in the selector drum; these must be renewed if bent or worn.

11 On MBX125 models, note that the excessive free play due to wear in the gearchange linkage may affect gear selection. Check the linkage components and renew any that are damaged or worn. At regular intervals (see Routine Maintenance) the linkage frame pivot should be dismantled for greasing to prevent stiffness and wear.

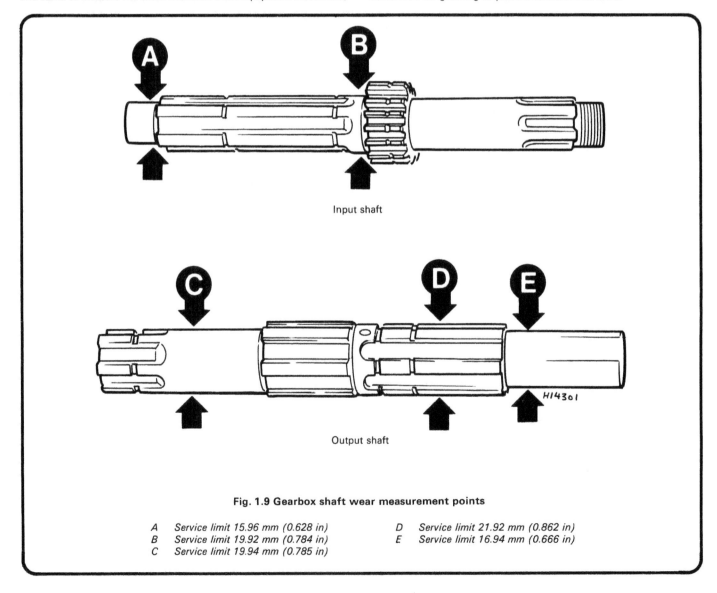

Fig. 1.9 Gearbox shaft wear measurement points

A	Service limit 15.96 mm (0.628 in)	D	Service limit 21.92 mm (0.862 in)
B	Service limit 19.92 mm (0.784 in)	E	Service limit 16.94 mm (0.666 in)
C	Service limit 19.94 mm (0.785 in)		

25.3 Check selector claw assembly for signs of wear or damage – note needle bearing

26 Gearbox shafts: dismantling and reassembly

Dismantling – general

1 The gearbox clusters should not be disturbed needlessly, and need only be stripped when careful examination of the whole assembly fails to reveal the cause of a problem, or where obvious damage, such as stripped or chipped teeth is discovered.

2 The input and output shaft components should be kept separate to avoid confusion during reassembly. Using circlip pliers, remove the circlip and plain washer which retain each part. As each item is removed, place it in order on a clean surface so that the reassembly sequence is obvious and the risk of parts being fitted the wrong way round or in the wrong sequence is avoided. Care should be exercised when removing circlips to avoid straining or bending them excessively. The clips must be opened just sufficiently to allow them to be slid off the shaft. Note that a loose or distorted circlip might fail in service, and any dubious items must be renewed as a precautionary measure. The same applies to worn or distorted thrust washers.

3 On the output shaft, note that the 4th gear pinion is located and retained by a splined thrust washer which is rotated so that its internal splines slide into the groove cut in the shaft splines; it is then held by a special locking washer. To remove the pinion, slide off the locking washer, rotate the thrust washer until its splines align with the shaft grooves, then pull it off the shaft.

4 The accompanying illustration shows how both clusters of the gearbox are assembled on their respective shafts. It is imperative that the gear clusters, including the thrust washers, are assembled in exactly the correct sequence, otherwise constant gear selection problems will occur. In order to eliminate the risk of misplacement, make rough sketches as the clusters are dismantled. Also strip and rebuild as soon as possible to reduce any confusion which might occur at a later date.

Reassembly – general

5 Having checked and renewed the gearbox components as required (see Section 25), reassemble each shaft, referring to the accompanying line drawing and photographs for guidance. Note that the manufacturer specifies a particular way in which the circlips are to be fitted. If a non-wire type circlip is examined closely, it can be seen that one surface has rounded edges and the other has sharply-cut square edges, this feature being a by-product of any stamped component such as a circlip or thrust washer. The manufacturer specifies that each circlip must be fitted with the sharp-edged surface facing away from any thrust received by the circlip. In the case of the gearbox shaft circlips this means that the rounded surface of any circlip must face towards the gear pinion that it secures. Furthermore, when a circlip is fitted to a splined shaft, the circlip ears must be positioned in the middle of one of the splines. These two simple precautions are specified to ensure that each circlip is as secure as is possible on its shaft and is best able to carry out its task. Ensure that the bearing surfaces of each component are liberally oiled before fitting.

6 If problems arise in identifying the various gear pinions which cannot be resolved by reference to the accompanying photographs and illustrations, the number of teeth on each pinion will identify them. Count the number of teeth on the pinion and compare this figure with that given in the Specifications Section of this Chapter, remembering that the output shaft pinions are listed first, followed by those on the input shaft. The problem of identification of the various components should not arise, however, if the instructions given in paragraph 2 of this Section are followed carefully.

7 When rebuilding the output shaft, note the special retaining arrangement of the 4th gear pinion. When the pinion has been refitted, slide the splined thrust washer down the shaft and rotate it until its internal splines are locked firmly in those of the shaft, then fit the locking washer to retain it.

8 Once any shaft has been rebuilt, tighten a rubber band over each end to prevent the loss of any component.

26.6a Input shaft is identified by the integral 1st gear pinion

26.6b 5th gear pinion is fitted with selector dogs facing to the left ...

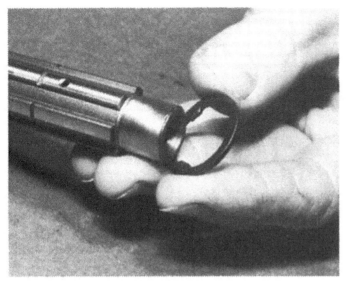

26.6c ... it is located by a splined thrust washer ...

26.6d ... and is retained by a circlip – note correct fitting

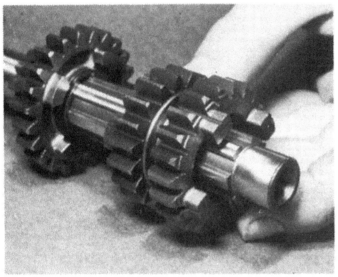

26.6e Identify 3rd and 4th gears by number of teeth to ensure that double pinion is correctly refitted

26.6f Install circlip in groove shown ...

26.6g ... followed by a splined thrust washer ...

26.6h ... and the 6th gear pinion

26.6i Finally fit the 2nd gear pinion ...

26.6j ... followed by the plain thrust washer

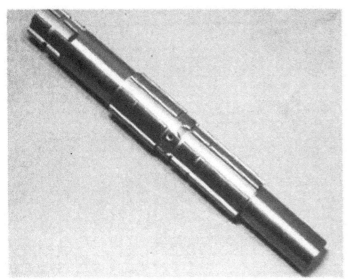

26.7a Take the bare output shaft ...

26.7b ... and fit the 4th gear pinion over its right-hand end, to rest against the shaft shoulder

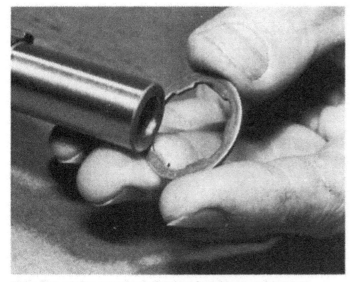

26.7c Splined thrust washer is fitted against 4th gear pinion and rotated to lock in shaft splines

26.7d Special locking washer is fitted as shown ...

26.7e ... to lock 4th gear pinion in place

26.7f Fit its centre bush and 3rd gear pinion as shown ...

26.7g ... followed by the splined thrust washer

26.7h Retain 3rd gear pinion assembly by fitting circlip as shown

26.7i Fit 5th gear pinion as shown – note position of selector fork groove

26.7j Place plain thrust washer over shaft end, followed by centre bush and 1st gear pinion

26.7k Finally fit the second plain thrust washer over shaft right-hand end

26.7l Fit 6th gear pinion over shaft left-hand end – note position of selector fork groove

26.7m Fit plain thrust washer against shaft shoulder ...

26.7n ... followed by its centre bush and the 2nd gear pinion

26.7o Finally fit plain thrust washer over shaft end – use rubber band to prevent loss of any component

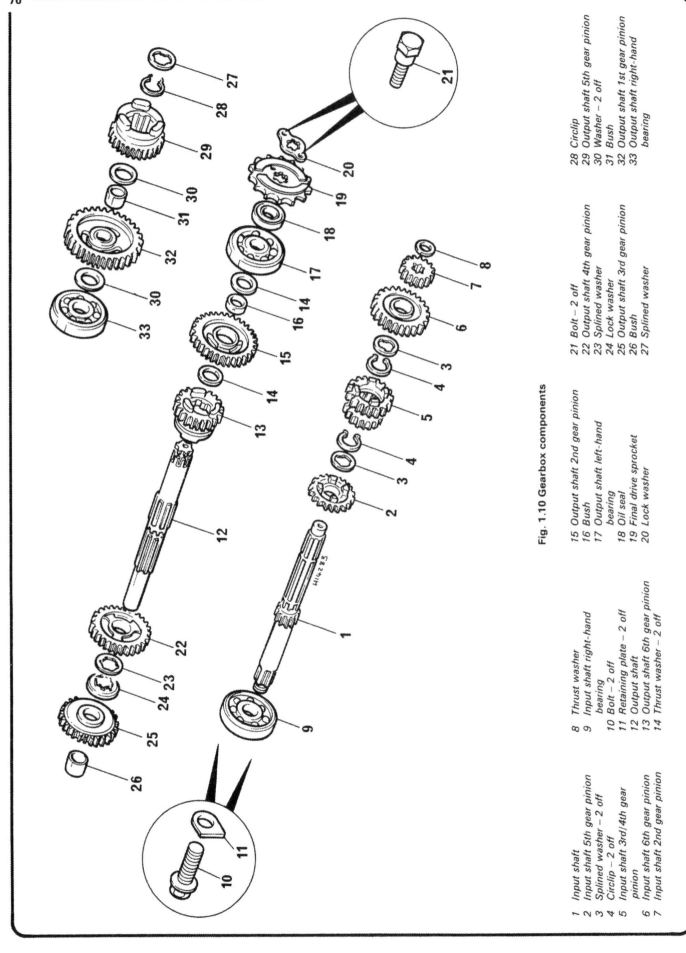

Fig. 1.10 Gearbox components

1   Input shaft
2   Input shaft 5th gear pinion
3   Splined washer – 2 off
4   Circlip – 2 off
5   Input shaft 3rd/4th gear
    pinion
6   Input shaft 6th gear pinion
7   Input shaft 2nd gear pinion

8   Thrust washer
9   Input shaft right-hand
    bearing
10 Bolt – 2 off
11 Retaining plate – 2 off
12 Output shaft
13 Output shaft 6th gear pinion
14 Thrust washer – 2 off

15 Output shaft 2nd gear pinion
16 Bush
17 Output shaft left-hand
    bearing
18 Oil seal
19 Final drive sprocket
20 Lock washer

21 Bolt – 2 off
22 Output shaft 4th gear pinion
23 Splined washer
24 Lock washer
25 Output shaft 3rd gear pinion
26 Bush
27 Splined washer

28 Circlip
29 Output shaft 5th gear pinion
30 Washer – 2 off
31 Bush
32 Output shaft 1st gear pinion
33 Output shaft right-hand
    bearing

27 Engine reassembly: general

1 Before reassembly of the engine/gear unit is commenced, the various component parts should be cleaned thoroughly and placed on a sheet of clean paper, close to the working area.

2 Make sure all traces of old gaskets have been removed and that the mating surfaces are clean and undamaged. Great care should be taken when removing old gasket compound not to damage the mating surface. Most gasket compounds can be softened using a suitable solvent such as methylated spirits, acetone or cellulose thinner. The type of solvent required will depend on the type of compound used. Gasket compound of the non-hardening type can be removed using a soft brass-wire brush of the type used for cleaning suede shoes. A considerable amount of scrubbing can take place without fear of harming the mating surfaces. Some difficulty may be encountered when attempting to remove gaskets of the self-vulcanising type, the use of which is becoming widespread, particularly as cylinder head and base gaskets. The gasket should be pared from the mating surface using a scalpel or a small chisel with a finely honed edge. Do not, however, resort to scraping with a sharp instrument unless necessary.

3 Gather together all the necessary tools and have available an oil can filled with clean engine oil, Make sure that all new gaskets and oil seals are to hand, also all replacement parts required. Nothing is more frustrating than having to stop in the middle of a reassembly sequence because a vital gasket or replacement has been overlooked. As a general rule each moving engine component should be lubricated thoroughly as it is fitted into position.

4 Make sure that the reassembly area is clean and that there is adequate working space. Refer to the torque and clearance settings wherever they are given. Many of the smaller bolts are easily sheared if overtightened. Always use the correct size screwdriver bit for the cross-head screws and never an ordinary screwdriver or punch. If the existing screws show evidence of maltreatment in the past, it is advisable to renew them as a complete set.

28 Reassembling the engine/gearbox unit: preparing the crankcases

1 At this stage the crankcase castings should be clean and dry with any damage, such as worn threads, repaired. If any bearings are to be refitted, the crankcase casting must be heated first as described in Section 14.

2 Place the heated casting on a wooden surface, fully supported around the bearing housing. Position the bearing on the casting, ensuring that it is absolutely square to its housing then tap it fully into place using a hammer and a tubular drift such as a socket spanner which bears only on the bearing outer race. Be careful to ensure that the bearing is kept absolutely square to its housing at all times.

3 Oil seals are fitted into a cold casing in a similar manner. Apply a thin smear of grease to the seal circumference to aid the task, then tap the seal into its housing using a hammer and a tubular drift which bears only on the hard outer edge of the seal, thus avoiding any risk of the seal being distorted. Tap each seal into place until its flat outer surface is just flush with the surrounding crankcase, or against its locating shoulder, where appropriate. Oil seals are fitted with the spring-loaded lip towards the liquid (or gas) being retained, ie with the manufacturer's marks or numbers facing outwards. Where double-lipped seals are employed, eg right-hand main bearing, use the marks or numbers to position each seal correctly if no notes were made on removal.

4 Where retaining plates are employed to secure bearings or oil seals, thoroughly degrease the threads of the mounting screws, apply a few drops of thread locking compound to them, and tighten them securely.

5 When all bearings and oil seals have been fitted and secured, lightly lubricate the bearings with clean engine oil and apply a thin smear of grease to the sealing lips of each seal.

6 Support the crankcase right-hand half on two wooden blocks placed on the work surface; there must be sufficient clearance to permit the crankshaft and gearbox components to be fitted.

28.2 Crankcase must be heated to permit easy refitting of bearings

28.3a Fitting oil seals – tubular drift must bear only on seal hard outer edge

28.3b If oil seals are refitted with shaft in place, protect seal lips from damage using insulating tape as shown

29 Reassembling the engine/gearbox unit: refitting the crankshaft and the gearbox components

1   Refit the rotor retaining nut to protect the crankshaft threaded end and insert the crankshaft as far as possible into its main bearing, using a smear of oil to ease the task. Align the connecting rod with the crankcase mouth, check that the crankshaft is square to the crankcase and support the flywheels at a point opposite the crankpin to prevent distortion while the crankshaft is driven home with a few firm blows from a soft-faced mallet. Do not risk damaging the crankshaft by using excessive force; if undue difficulty is encountered, take the assembly to a Honda Service Agent for the crankshaft to be drawn into place using the correct service tool.
2   When the crankshaft is fitted, remove the protecting nut and check that the crankshaft revolves easily with no trace of distortion.
3   Fit the two gear clusters together, ensuring that all pinions are correctly meshed, remove the retaining rubber bands from the shaft right-hand end(s) and insert them as a single unit into the crankcase. Insert the selector drum and rotate it to the neutral position, with the indicator switch contact in the 3 o'clock position.
4   Refit the output shaft selector forks to their respective pinion grooves, engage their guide pins on the selector drum then oil the fork shaft and push it through the fork bores and into the crankcase. Similarly, refit the single input shaft fork and its shaft.
5   To identify each fork use the marks made on dismantling. If none were made it was found that the manufacturer had provided marks in the form of a letter cast into one surface of each fork. The fork marked 'R' is fitted to the output shaft 5th gear pinion with the letter facing to the right, that marked 'L' is fitted to the output shaft 6th gear pinion with the letter facing to the left and the fork marked 'C' is fitted to the input shaft 3rd/4th gear pinion with its letter facing to the left.
6   When all components are fitted, check that the selector drum is in the neutral position and that both shafts are free to rotate, then rotate the drum to check that all six gears can be selected with relative ease. Return to the neutral position and lubricate thoroughly all bearings and bearing surfaces.
7   Lubricate thoroughly its bearings and refit the balancer shaft, tapping it home until its bearing outer race is flush with the crankcase right-hand surface. Ensuring that it is correctly rebuilt as described in Section 24, refit the kickstart shaft assembly. If it was disturbed, refit the ratchet guide plate as shown in the accompanying photograph. Apply thread locking compound to their threads and refit the two

retaining bolts in the positions shown, tightening each to a torque setting of 1.0 – 1.6 kgf m (7 – 11.5 lbf ft).
8   Fit the kickstart pinion into the crankcase bore from the right-hand side, ensuring that its protruding boss faces to the left; hold it in position then insert the shaft assembly into the pinion so that the ratchet stop engages with the guide plate as shown in the accompanying photograph. Use a heavy pair of pliers to bring the long hooked end of the return spring around until it can be engaged on the crankcase lug. Refit temporarily the lever and check the kickstart operation, then refit the thrust washer to the shaft right-hand end and retain it with a rubber band so that it cannot be lost.
9   The water pump shaft (except MTX125 RW-H and L models) is best refitted in the left-hand crankcase half. If the pump bearing and shaft are in one piece they should be fitted into the crankcase from its left-hand side as described in the previous Section. Insert the drive pin into the shaft and fit the driven gear with its raised boss facing to the left. Refit the thrust washer and the circlip to retain it, then check that the shaft and the drive gear/crankcase breather assembly are free to rotate.

29.1 Crankshaft is easily distorted – take care not to use excessive force on refitting

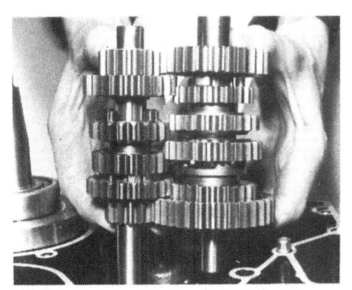

29.3a Fit gear clusters together and insert as single unit – take care not to lose thrust washers

29.3b Fit selector drum and rotate to neutral position

29.4a Position selector forks on pinions and drum, then oil shaft before refitting ...

29.4b ... input shaft fork is fitted in the same way

29.5 Selector forks may be identified by cast-in marks

29.7a Lubricate balancer shaft bearings before refitting

29.7b Kickstart ratchet guide plate correctly installed – note position of different bolts

29.8a Refitting the kickstart shaft assembly – fit pinion into crankcase as shown ...

29.8b ... and insert shaft from left-hand side so that ratchet stop engages as shown

29.8c Hook return spring end over crankcase lug – check it is secure

29.8d Fit thrust washer over shaft right-hand end – retain wth rubber band

29.9 So that drive gears can be correctly meshed, water pump shaft is best fitted into left-hand crankcase half – except MTX125 RW-H and L models

30 Reassembling the engine/gearbox unit: joining the crankcase halves

1   Press both locating dowel pins into their recesses in the right-hand crankcase gasket surface, then fit a new gasket. Make a final check that all components are in position and that all bearings and bearing surfaces are lubricated. Remove the rubber bands (if used) from the gearbox shafts.
2   Ensuring that the water pump shaft (except MTX125 RW-H and L) is inserted correctly into the balancer shaft, lower the left-hand crankcase half into position, using firm hand pressure only to push it home. It may be necessary to give a few gentle taps with a soft-faced mallet to drive the casing fully into place. Do not use excessive force, instead be careful to check that all shafts and dowels are correctly fitted and accurately aligned, and that the crankcase halves are exactly square to each other. If necessary, pull away the upper crankcase half to rectify the problem before starting again.
3   When the two halves have joined correctly and without strain, refit the crankcase retaining screws using the cardboard template to position each screw correctly. Working in a diagonal sequence from the centre outwards, progressively tighten the screws until all are securely and evenly fastened; the recommended torque setting is 0.8 – 1.2 kgf m (6 – 9 lbf ft).
4   Trim away the surplus gasket from the crankcase mouth, then check the free running and operation of the crankshaft and gearbox components. If a particular shaft is stiff to rotate, a smart tap on each end using a soft-faced mallet, will centralise the shaft in its bearing. If this does not work, or if any other problem is encountered, the crankcases must be separated again to find and rectify the fault. Pack clean rag into the crankcase mouth to prevent the entry of dirt, then refit the drain plug, if removed, tightening it to the specified torque setting.

30.1 Fit new gasket over two locating dowels (arrowed). Water pump shaft (except MTX125 RW-H, L) shown in position for reference only – fit as described in text

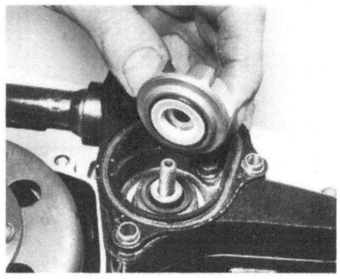

31.3 Do not omit mechanical seal or copper washer when refitting water pump impeller

31 Reassembling the engine/gearbox unit: refitting the water pump

1   Having fitted a new seal as described in Section 28, fit a new sealing O-ring to the pump seal collar and apply a smear of grease to the O-ring and to the seal lips before pressing the collar squarely into its bore in the crankcase.
2   Having checked the mechanical seal and renewed it if necessary, as described in Chapter 2, fit a new gasket to the pump body and refit it to the crankcase ensuring that the crankcase breather hose and coolant pipes are correctly installed. If the crankcase left-hand cover is not to be refitted yet, install temporarily the second (rear) pump body mounting bolt to ensure that the body is correctly aligned, and tighten the front mounting bolt to a torque setting of 0.8 – 1.2 kgf m (6 – 9 lbf ft).
3   Fit the copper washer to the pump shaft and screw the impeller **anti-clockwise** on to the shaft; remember that a **left-hand** thread is employed. Use the method employed on dismantling to lock the pump shaft and tighten the impeller to a torque setting of 1.3 – 1.7 kgf m (9.5 – 12 lbf ft). Check that it is free to rotate.
4   Fit the two locating dowels and two new sealing O-rings to their recesses in the body gasket surface, then refit the pump cover and tighten its three mounting bolts to a torque setting of 0.8 – 1.2 kgf m (6 – 9 lbf ft).

31.4 Refitting water pump cover – note two locating dowels and two sealing O-rings

32 Reassembling the engine/gearbox unit: refitting the alternator

1   If it was removed, fit a new sealing washer to the neutral indicator switch and screw it into the crankcase, tightening it carefully.
2   Ensuring that their leads are correctly routed refit first the pulser coil, then the alternator stator. Check that the wire sealing rubber grommets are pressed correctly into their cutouts in the crankcase wall. Refit the pulser coil wiring clamp and tighten its bolt to a torque setting of 0.8 – 1.2 kgf m (6 – 9 lbf ft), then tighten the pulser coil mounting bolt to the same torque setting, having applied a few drops of thread locking compound to their threads. Refit and tighten securely with an impact driver the three stator mounting screws; apply thread locking compound to their threads also. Do not forget to connect the neutral switch.
3   Degrease the rotor and crankshaft tapers. Insert the Woodruff key into the crankshaft and push the rotor over it. Gently tap the rotor centre with a soft-faced hammer to seat it and fit the washer and nut. Lock the crankshaft and tighten the nut to a torque setting of 6.0 – 7.0 kgf m (43 – 50.5 lbf ft).

32.2 Do not forget to refit and connect neutral indicator switch

32.3a Degrease crankshaft taper and insert Woodruff key ...

32.3b ... align rotor keyway with key – check rotor is securely seated ...

32.3c ... before refitting washer and retaining nut

32.3d Lock crankshaft and tighten nut to specified torque setting

33 Reassembling the engine/gearbox unit: refitting the gear selector mechanism external components and the ancillary drive components

1   Insert the five selector pins into the selector detent cam and refit the cam to the selector drum, ensuring that the cutout in the cam engages with the key or locating projection on the selector drum. Refit the pin retaining plate and the circlip.

2   Place its thrust washer on the crankcase boss and refit the detent roller arm, ensuring that its return spring is correctly fitted. Apply thread locking compound to its threads and refit the shouldered pivot bolt, tightening it to a torque setting of 0.8 – 1.2 kgf m (6 – 8 lbf ft). Check that the arm is free to move against spring pressure and that the roller is engaged correctly on the selector detent cam.

3   Check that its return spring is fitted correctly, as shown in the accompanying photograph and smear grease over its splines to protect the oil seal lips, then insert the gearchange shaft. Fit the reduction gear pivot post into its crankcase recess, ensuring that it fits between the return spring ends. Fit the thrust washer to the claw arm pivot on the gearchange shaft. Aligning it so that its centre tooth engages correctly with that of the gearchange shaft, refit the reduction gear and its thick spacer to the pivot post. Finally, fit the claw arm assembly to the

gearchange shaft ensuring that its teeth are engaged correctly with the reduction gear centre tooth and that the claw arm is aligned with the selector pins. Refit temporarily the gearchange pedal and check that all gears can be selected.

4   On MTX200 models only, grease thoroughly the ATAC shaft as described in Chapter 3 and refit it. It may be necessary to rotate the crankshaft so that the shaft gear teeth can be slid past those of the primary drive gear and into engagement with those of the ATAC drive gear. Do not forget the thrust washer(s) if fitted.

5   Refitting its thrust washer to the shaft and ensuring that the Woodruff key is firmly inserted into the shaft keyway, refit the balancer/water pump drive idler shaft; the Woodruff key must engage with the cutout in the crankcase breather/water pump drive gear assembly. Check very carefully that this is correct, withdrawing and refitting the shaft if necessary. As the anti-backlash gears begin to mesh with the balancer shaft gear, use a small pair of pliers to align both sets of anti-backlash gear teeth so that the shaft can correctly engage both driven components at the same time. If the Woodruff key is dislodged and falls clear, the crankcases must be separated again to recover it.

6   Check that the balancer shaft, water pump shaft (except MTX125 RW-H, L) and the idler shaft are correctly fitted and that they are free to rotate easily and smoothly. Lubricating them thoroughly, fit the kickstart idler gear and its bush to the output shaft.

33.1a Insert selector pins into detent cam

33.1b Detent cam is located on selector drum by key shown

33.2a Refitting detent roller arm – place thrust washer on crankcase boss ...

33.2b ... then fit arm and return spring as shown – check arm is free to move after pivot bolt has been correctly tightened

33.2c Roller arm should engage as shown with detent cam – note pin retaining plate and circlip on detent cam

33.3a Gearchange shaft return spring free ends should fit as shown

33.3b Fit gearchange shaft assembly into crankcase ...

33.3c ... then insert reduction gear pivot post as shown

33.3d Fit thrust washer as shown to gearchange shaft

33.3e Align teeth carefully on refitting reduction gear

33.3f Do not omit reduction gear spacer

33.3g Claw arm assembly is fitted as shown – ensure teeth are correctly aligned

33.5 Align teeth of anti-backlash gear so that they will engage with those of balancer shaft gear

33.6a Kickstart idler gear centre bush is fitted as shown ...

33.6b ... and should be oiled before idler gear is refitted

34 Reassembling the engine/gearbox unit: refitting the clutch and primary drive pinion

1   Grease the lips of the crankshaft right-hand oil seal and refit the spacer to the crankshaft, taking care not to damage the seal, then refit the ATAC shaft drive gear (MTX200 models only) and the primary drive gear. As the primary drive gear is refitted, align the punch mark on its right-hand face with the punch mark on one of the crankshaft splines.

2   If the primary drive gear retaining nut lock washer is of the Belville type it is refitted with its convex surface outwards; this surface is usually marked 'OUTSIDE' to ensure that there can be no mistake. If the (lock) washer is just a standard type plain washer there is no particular way to fit it.

3   Fit the washer and nut to the crankshaft, lock the crankshaft by the method used on dismantling and tighten the nut to a torque setting of 6.0 – 7.0 kgf m (43 – 50.5 lbf ft). Fit the first thrust washer and the clutch outer drum bearing sleeve to the input shaft and oil the sleeve.

4   To prevent excessive vibration, the balancer shaft must be synchronized exactly with the crankshaft. Since the balance shaft is driven via the clutch outer drum, both shafts must be kept in perfect alignment with index marks cast on the crankcase wall when the outer drum is refitted. Refer to the accompanying photographs and illustration. Rotate the crankshaft and balancer shafts until the punch marks on the crankshaft and primary drive gear and the punch mark or scribed line on the balancer shaft gear align with their respective index marks. If possible, lock the shafts so that the marks are in exact alignment.

5   Using a pair of pliers align both sets of teeth of the anti-backlash gear at the point of contact with the clutch outer drum teeth so that the outer drum can be slid into place without disturbing the balancer shaft setting. Check again that the marks are aligned and slide the outer drum into place so that its teeth mesh correctly with those of the primary drive gear, the idler shaft and the kickstart idler gear.

6   Check carefully that the balancer timing marks are correctly aligned, removing and again refitting the outer drum if necessary, then fit the second thrust washer to the input shaft.

7   For ease of assembly, the clutch pressure plate, the clutch plain and friction plates, and the clutch centre should be assembled separately and offered up as a single unit. Place the clutch centre upside down on the work surface. If new friction plates are to be fitted, each should be coated with a light film of engine oil before assembly. On MTX200 RW-D models only, first fit the thick judder spring seat, then the spring itself, and the single friction plate which has the larger inside diameter so that it fits around the judder spring. Fit the remaining friction and plain plates alternately, as follows. On all other models, fit first a friction plate, then continue the assembly fitting plain and friction plates alternately, to finish with a friction plate. Fit the clutch pressure plate, passing the four projecting pillars through the corresponding holes in the clutch centre, then align the projecting tongues on the friction plates to ensure that the assembly will slide easily into the clutch outer drum. Fit the whole clutch centre/pressure plate assembly into the clutch outer drum.

8   Fit the clutch centre retaining nut (lock) washer as described in paragraph 2 of this Section, followed by the retaining nut itself. Lock the clutch centre by the method used on dismantling and tighten the nut to a torque setting of 6.0 – 7.0 kgf m (43 – 50.5 lbf ft).

9   Fit the clutch springs to their pillars and refit the clutch lifter plate. Tighten the four bolts evenly in a diagonal sequence and in two or three stages so that spring pressure is evenly applied.

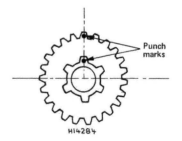

H14284

Fig. 1.11 Primary drive gear/crankshaft alignment punch marks

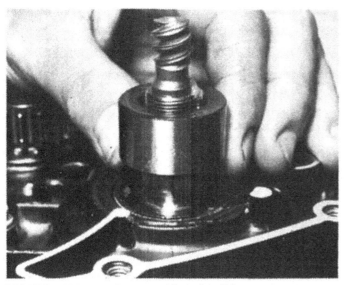

34.1a Take care not to damage seal lips when fitting spacer -- use grease to assist fitting

34.1b When fitting primary drive gear, align punch mark with mark on crankshaft spline (arrowed)

34.2 Different types of lock washer may be used – fit as described in text

34.3a Lock crankshaft and tighten primary drive gear retaining nut to specified torque setting

34.3b Fit narrower thrust washer against input shaft bearing ...

34.3c ... followed by clutch outer drum bearing sleeve – oil liberally

34.4a Crankshaft and primary drive gear punch marks (aligned on fitting gear) should align with crankcase index mark (arrowed)

34.4b Balancer shaft mark should align with crankcase index mark (arrowed)

34.5 Ensure balancer shaft timing is not disturbed on refitting clutch outer drum

34.6 Wider thrust washer is fitted against clutch outer drum, as shown

34.7a Starting with a friction plate, place clutch plates over inverted clutch centre ...

34.7b ... building them up alternately until all are fitted

34.7c Align friction plate tangs and fit pressure plate

34.8a Ensure clutch centre nut and lock washer are correctly fitted – see text

34.8b Lock clutch centre as shown and tighten retaining nut to specified torque setting

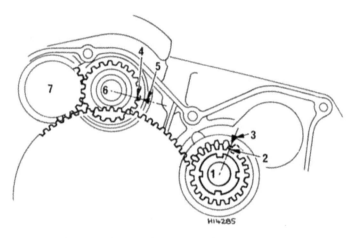

Fig. 1.12 Crankshaft and balancer shaft alignment marks

1 Primary drive gear	5 Index mark
2 Punch mark	6 Balancer gear
3 Index mark	7 Idler gear
4 Punch mark	

35 Reassembling the engine/gearbox unit: refitting the right-hand crankcase cover

1 Check all components within the cover are properly fitted and lubricated. Grease the splined end of the kickstart shaft and the lip of the cover oil seal. If it was removed, check that the clutch release mechanism is correctly refitted and greased, and that the tachometer drive gear is correctly fitted and lubricated.
2 Degrease the cover and crankcase mating surfaces, press the two locating dowels into the crankcase and locate the new gasket over them. Push the cover over the dowels, tapping it lightly with a soft-faced hammer to seat it properly.
3 Fit each screw into its previously noted position. Tighten the screws evenly whilst working in a diagonal sequence to a torque setting of 0.8 – 1.2 kgf m (6 – 9 lbf ft). Do not forget to fit the clutch cable bracket, or to check that the clutch release arm is operating smoothly.

4 On MTX200 models only, renew its sealing O-ring. grease the components as described in Chapter 3 and refit the ATAC operating spindle, Remove the filler plug and use a torch to check that the groove in the pulley-type wheel at the end of the spindle engages correctly with the central raised rib of the operating shaft rack. Tighten the mounting nut to a torque setting of 1.8 – 2.0 kgf m (13 – 14.5 lbf ft).
5 On all models, renew its sealing O-ring, oil its drive shaft and fit the oil pump. Tighten the mounting bolt to a torque setting of 0.8 – 1.2 kgf m (6 – 9 lbf ft).

36 Reassembling the engine/gearbox unit: refitting the piston, cylinder barrel and head

1 The piston rings are refitted using the same technique as on removal. Do not forget to fit the expander to the bottom piston ring groove before refitting that ring. The top surface of each ring is

indicated by the letter 'N', 'R' or 'T' (depending on the ring manufacturer) stamped near the ring end gap; ensure that both rings have the same identifying letter. If the rings are being re-used and these marks cannot be seen the bottom (plain) ring, must be installed the original way up using the wear marks for identification. The top (Keystone) ring has a slight inwards taper on its top surface; check that it is fitted to the top groove only and with its tapered surface upwards.

2   Rotate both rings so that their end gaps align with the locating peg in each ring groove to stop the rings rotating and catching in one of the ports. See the accompanying photograph.

3   Lubricate the big-end bearing and main bearings, bring the connecting rod to the top of its stroke, then pack the crankcase mouth with clean rag. Lubricate and fit the small-end bearing. With the 'IN' mark cast on the piston crown facing to the rear, place the piston over the rod and fit the gudgeon pin. If necessary, warm the piston to aid fitting.

4   Retain the gudgeon pin with **new** circlips. Check each clip is correctly located, as shown in the accompanying photograph; if allowed to work loose it will cause serious damage.

5   Check the piston rings are still correctly fitted. Clean the barrel and crankcase mating surfaces and fit the new base gasket with the moulded rubber seam against the crankcase. Lubricate the piston rings and cylinder bore. Lower the barrel into position and ease the piston into the bore, carefully squeezing the ring ends together. Do not use excessive force.

6   Remove the rag from the crankcase mouth and push the barrel down on to the crankcase, then connect the coolant hoses or pipes while there is freedom to manoeuvre them into place.

7   The two cylinder barrel retaining nuts inside the water jacket and the cylinder head retaining nuts are likely to suffer from corrosion. It is recommended that an anti-seize compound such as Copaslip is applied to their threads so that they can be removed easily in the future.

8   Refit the four cylinder barrel nuts and tighten them evenly and in a diagonal sequence to a final torque setting of 2.0 – 2.4 kgf m (14.5 – 17 lbf ft). Check that the coolant hoses or pipes are correctly connected and tighten or refit as applicable the clamps or clamping bolts and screws.

9   On MTX200 models only, refit the rubber cover over the ATAC operating spindle then assemble the spring and both parts of the connecting arm (ensuring that the longer arm is uppermost) and fit them to the spindle. Ensure that the squared hole in the shorter (guide plate) arm engages correctly with the spindle cutaway section and that the pin in the end of the longer arm projects downwards between the washer and valve cotter retainer on the valve stem; the spring should hold the two arms securely together. Refit the plain washer (where fitted), the collar and the nut; tighten the nut securely. Do not fit the rubber cover over the valve stem until the engine has been started and the system checked.

10   On all models, clean the head and barrel gasket surfaces, fit a new gasket over the studs and refit the cylinder head. Apply anti-seize compound to the stud threads, refit the five retaining nuts and tighten them evenly, in two or three stages, and in a diagonal sequence to a torque setting of 2.0 – 2.4 kgf m (14.5 – 17 lbf ft). Remember that the head is a thin and delicate casting which may be easily distorted or even cracked if not correctly fastened.

11   Refit the two dowels over the studs in the cylinder barrel water jacket, fit the seal around the cylinder head spark plug boss and the large seal to the cylinder head cover; these seals are not subjected to very great compression and may be re-used if serviceable. Refit the cylinder head cover and its two retaining nuts and washers. Tighten the nuts to a torque setting of 0.8 – 1.2 kgf m (6 – 9 lbf ft). Check that the spark plug electrodes are clean and correctly gapped, then refit the plug, as described in Routine Maintenance.

12   Fit a new gasket to the inlet port and refit the reed valve assembly, followed by a second gasket and the inlet stub. Tighten the four retaining bolts evenly to a torque setting of 0.8 – 1.2 kgf m (6 – 9 lbf ft). The oil feed pipe may be unplugged and connected to the inlet stub union, but remember that it will be necessary to carry out the bleeding procedure if air bubbles have been allowed into the pipe.

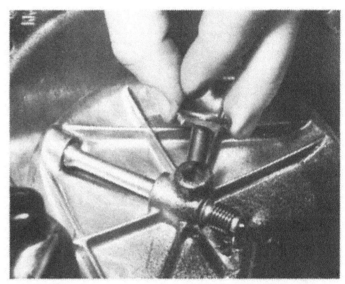

35.1 Grease clutch pushrod and press firmly into cover recess

35.2 Fit new cover gasket over two locating dowels (arrowed)

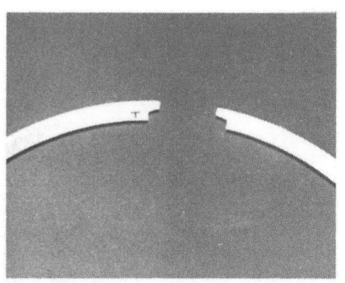

36.1 Letter stamped near end gap marks ring top surface – also both rings must have same letter

36.2 Ensure ring end gaps align with piston locating pegs as shown

36.3a Lubricate thoroughly all bearings -- note rag packing crankcase mouth

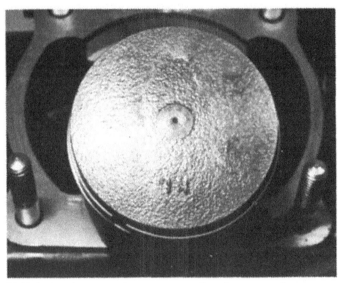

36.3b Use mark cast on piston crown to align piston correctly on refitting

36.4 Gudgeon pin retaining circlips are a special type – will fit correctly only one way

36.5 Be careful to fit cylinder base gasket the correct way up

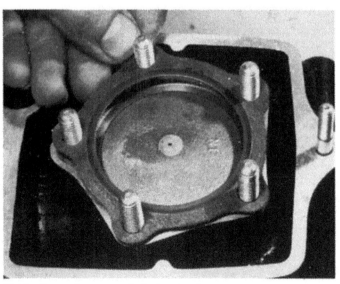

36.10a Always fit a new cylinder head gasket ...

36.10b ... then fit the cylinder head

36.10c Apply anti-seize compound to stud threads ...

36.10d ... and tighten retaining nuts to specified torque setting

36.11 Refit the cylinder head cover – note two seals and two locating dowels

36.12 Always use new gaskets when refitting reed valve and inlet stub

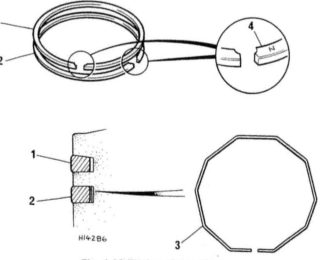

Fig. 1.13 Fitting piston rings

1   Top ring
2   Second ring
3   Expander
4   Ring identification mark

37 Refitting the engine/gearbox unit into the frame

1 Check that no part has been omitted during reassembly. Prepare the machine and ease the engine into position. Align the engine and fit the mounting bolts and nuts; tighten these to the specified torque settings, having ensured that the engine is seated correctly, that the spacer(s) have been refitted in their correct positions (where applicable) and that no components are trapped.

2 Refit the cylinder head cover (if removed) as described in the previous Section, check the spark plug as described in Routine Maintenance and refit it. Connect the suppressor cap to the spark plug and wire to the temperature gauge sender unit. On MTX 125/200 models, refit the chain tensioner assembly and the rear brake pedal assembly.

3 Refit the carburettor to the intake stub, check it is vertical and tighten the clamp then refit the air filter hose and tighten its clamp. Insert the throttle slide assembly, ensuring that it fits correctly, then tighten the carburettor top. Refit the oil pump cable to the pump. Working as described in Routine Maintenance, check the throttle cable and the pump cable adjustment. Connect the oil tank/pump feed pipe – again ensuring that it is correctly routed, and bleed any air from the feed pipe and pump using the bleed screw, as described in Chapter 3. On MBX125 models, route the crankcase breather and float bowl drain tubes down the rear end of the crankcase and secure them with the clamp provided. On MTX125/200 models, route the crankcase breather tube up to its union on the air filter casing and connect it; the float bowl drain tube is routed as described for MBX125 models.

4 Route the main generator lead from the crankcase up to the frame, connect it again to the main loom and secure it with any clamps or ties provided. Refit its rubber cover to the connector. Referring to Chapter 2, refit the radiator, connect all the coolant hoses and/or pipes and refill the system with coolant, but leave the radiator cap off as a reminder that the level must be checked. Ensure that both drain plugs are fastened to the specified torque setting of 0.8 – 1.2 kgf m (6 – 9 lbf ft).

5 Engage the gearbox sprocket on the chain and fit the sprocket to the output shaft end. Slide the locking plate along the shaft and rotate it so that its bolt holes align with those of the sprocket and its internal splines are locked in the groove cut in the shaft splines. Apply thread locking compound to their threads and refit the two retaining bolts; apply the rear brake hard to prevent rotation while they are tightened securely. Working as described in Routine Maintenance check, and adjust if necessary, the chain tension, rear brake adjustment and stop lamp rear switch setting.

6 Refit the crankcase left-hand cover, tightening its retaining bolts to a torque setting of 0.8 – 1.2 kgf m (6 – 9 lbf ft). Using the marks made on dismantling to align it correctly, refit the gearchange pedal or linkage front arm (as appropriate), then refit its pinch bolt, tightening it to a torque setting of 0.8 – 1.2 kgf m (6 – 9 lbf ft). On MBX125 models, check that the angles formed between the linkage front and rear arms and the adjusting rod are at 90° when the pedal is at the required height. If adjustment is required, slacken the adjusting rod locknuts and remove the front arm pinch bolt so that the arm can be pulled off the gearchange shaft splines and rotated; adjustment is a matter of rotating the rod as necessary to lengthen or shorten it while rotating the front arm about the shaft splines to preserve the 90° angles. When both angles are correct and the pedal is at the required height, press the front arm on to the shaft and secure its pinch bolt as described above, then tighten securely the rod locknuts.

7 Fit the kickstart to its shaft, using the marks made on dismantling to ensure that it is refitted in the same position, then refit its pinch bolt, tightening it to a torque setting of 2.0 – 2.4 kgf m (14.5 – 17 lbf ft). Connect the clutch cable to its adjuster bracket and operating lever then adjust the cable as described in Routine Maintenance. Insert the tachometer cable into its housing; it may be necessary to turn the engine over on the kickstart to engage the drive and cable. Tighten securely the cable retaining screw.

8 Refit the exhaust system. Use grease to stick a new gasket to the exhaust port and fit the mountings loosely. Tighten the exhaust port fasteners first, then the remaining mountings. On MTX125/200 models, do not forget to check the tubular seal between the exhaust pipe and the tailpipe/silencer; this must be renewed if damaged, split or broken. Tighten all exhaust mounting bolts and nuts to the specified torque settings.

9 Connect the battery to the main loom, checking its terminals and the vent hose as described in Routine Maintenance. Check that the transmission oil drain plug is fastened to the specified torque setting of 2.0 – 2.5 kgf m (14.5 – 18 lbf ft) and pour in 600 cc (1.1 pint) of SAE 10W30 SE or SF engine oil; if this amount cannot be measured accurately use the level plug as described in Routine Maintenance but note that the level will drop after the engine has been run for the first time as the oil is distributed around the various components and will therefore require careful re-checking. Refit the filler (and level) plugs.

10 Refit the fuel tank as described in Chapter 3, turn on the tap and check for any fuel leaks, which must be cured immediately. On MBX 125 models, refit the radiator shroud, the belly fairing and the right-hand fuel tank panel. On MTX125/200 models refit the crankcase bashplate, tightening its mounting bolts to a torque setting of 0.8 – 1.2 kgf m (6 – 9 lbf ft).

11 Make a final check that all components have been refitted and that all are correctly adjusted and securely fastened; the exceptions are the components of the ATAC, lubrication and cooling systems which are still to receive a final check before the machine is ready for use.

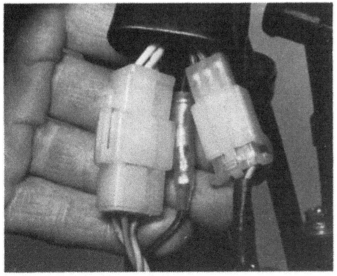

37.4a Connect generator wires to main loom

37.4b MBX125 – do not fully tighten radiator bottom mountings until radiator shroud and belly fairing are refitted

37.5 Refit gearbox sprocket and locking plate – tighten retaining bolts securely

37.6a MBX125 – Use link rod adjuster to ensure that linkage angles ...

37.6b ... are correctly aligned – see text

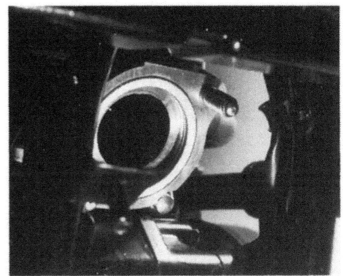

37.8 Always use a new gasket to prevent exhaust leaks

38 Starting and running the rebuilt engine

1   Attempt to start the engine using the usual procedure adopted for a cold engine. Do not be disillisioned if there is no sign of life initially. A certain amount of perseverance may prove necessary to coax the engine into activity even if new parts have not been fitted. Should the engine persist in not starting, check that the spark plug has not become fouled by the oil used during re-assembly. Failing this go through the fault finding charts and work out what the problem is methodically.

2   When the engine does start, keep it running as slowly as possible to allow the oil to circulate. Open the choke as soon as the engine will run without it. During the initial running, a certain amount of smoke may be in evidence due to the oil used in the reassembly sequence being burnt away. The resulting smoke should gradually subside.

3   Check the engine for blowing gaskets and coolant or oil leaks. Before using the machine on the road, check that all the gears select properly, and that the controls function correctly. As soon as the engine has warmed up to normal operating temperature and is running smoothly stop it and check the following.

4   The coolant level should drop as the engine runs, especially as soon as the thermostat opens, as the coolant is distributed around the various passages of the system. Trapped air will be expelled in the form of bubbles and must be allowed to escape before the radiator cap is refitted. As soon as the temperature gauge has reached its normal level and the level has stabilised with no more air bubbles appearing, stop the engine and top the coolant up to the bottom of the radiator filler neck, then refit and tighten securely the filler cap. If necessary, top the level in the expansion tank up to the 'Full' or 'F' mark.

5   While the coolant level is being checked, carry out the bleeding procedure described in Chapter 3 to remove air bubbles from the oil pump/inlet stub feed pipe by again disconnecting it. When all air has been removed reconnect the pipe to the inlet stub union and secure it with its clip. Mop up any spilled oil and check that the pipes are correctly routed and properly secured, then refit the oil pump cover.

6   On MTX200 models only, check the operation of the ATAC system as described in Chapter 3, before refitting the rubber cover and tightening its clamp.

7   It may be necessary to make some adjustment to the carburettor settings as described in Routine Maintenance and/or Chapter 3.

8   When all other checks have been completed, stop the engine and check the transmission oil level as described in Routine Maintenance; add oil as necessary to achieve the correct level.

9   Finally, refit all remaining components such as radiator shroud(s), side panels and the seat.

39 Taking the rebuilt machine on the road

1   Any rebuilt machine will need time to settle down, even if parts have been replaced in their original order. For this reason it is highly advisable to treat the machine gently for the first few miles to ensure oil has circulated throughout the lubrication system and that new parts fitted have begun to bed down.

2   Even greater care is necessary if the engine has been rebored or if a new crankshaft has been fitted. In the case of a rebore, the engine will have to be run in again, as if the machine were new. This means greater use of the gearbox and a restraining hand on the throttle until at least 500 miles have been covered. There is no point in keeping to any set speed limit; the main requirement is to keep a light loading on the engine and to gradually work up performance until the 500 mile mark is reached. These recommendations can be lessened to an extent when only a new crankshaft is fitted. Experience is the best guide since it is easy to tell when an engine is running freely.

3   Remember that a good seal between the piston and the cylinder barrel is essential for the correct functioning of the engine. A rebored two-stroke engine will require more careful running-in, over a long period, than its four-stroke counterpart. There is a far greater risk of engine seizure during the first hundred miles if the engine is permitted to work hard.

4   If at any time a lubrication failure is suspected, stop the engine immediately and investigate the cause. If an engine is run without oil, even for a short period, irreparable engine damage is inevitable.

5   Do not on any account add oil to the petrol under the mistaken belief that a little extra oil will improve the engine lubrication. Apart from creating excess smoke, the addition of oil will make the mixture much weaker, with the consequent risk of overheating and engine seizure. The oil pump alone should provide full engine lubrication.

6   Do not tamper with the exhaust system or run the engine without the baffle fitted to the silencer. Unwarranted changes in the exhaust system will have a marked effect on engine performance, invariably for the worse. The same advice applies to dispensing with the air cleaner or the air cleaner element.

7   When the initial run has been completed allow the engine unit to cool and then check all the fittings and fasteners for security. Re-adjust any controls which may have settled down during initial use and re-check the coolant level in the radiator and the transmission oil level.

Chapter 2 Cooling system

Contents

Specifications

Unless otherwise stated, information applies to all models

Coolant

Mixture type	Distilled water with corrosion-inhibited ethylene glycol antifreeze
Standard recommended mixture ratio	30% antifreeze: 70% distilled water
Total capacity approx:	
125 models	950 cc (1.67 pint)
200 models	1030 cc (1.81 pint)
Mixture ratios recommended at temperatures of down to:	
20% antifreeze: 80% distilled water	–9°C (16°F)
30% antifreeze: 70% distilled water	–16°C (3°F)
40% antifreeze: 60% distilled water	–25°C (–13°F)
50% antifreeze: 50% distilled water	–37°C (–35°F)
55% antifreeze: 45% distilled water	–44.5°C (–48°F)

Thermostat

Opens at	69.5 – 72.5°C (157–162.5°C)
Fully open at	80°C (176°F)
Valve minimum lift – fully open	4.5 mm (0.1772 in)

Radiator

Cap valve opening pressure	0.9 kg/cm² (12.8 psi)
Tolerance	0.75 – 1.05 kg/cm² (10.7 – 14.9 psi)
Radiator core testing pressure	0.90 ± 0.15 kg/cm² (12.8 ± 2.1 psi)

Torque settings

Component	kgf m	lbf ft
Temperature gauge sender unit	0.8 – 1.4	6 – 10
Water pump cover and body bolts	0.8 – 1.2	6 – 9
Water pump impeller	1.3 – 1.7	9.5 – 12
Coolant drain plugs	0.8 – 1.2	6 – 9
Radiator mounting bolts:		
MBX125	0.9 – 1.3	6.5 – 9.5
MTX125/200	0.8 – 1.2	6 – 9
Radiator cover/belly fairing mounting nuts – MBX125	0.8 – 1.2	6 – 9
Belly fairing mounting screws – MBX125	0.8 – 1.1	6 – 8
Expansion tank mounting bolt:		
MBX125	0.9 – 1.3	6.5 – 9.5
MTX125/200	0.8 – 1.2	6 – 9

1 General description

The liquid cooling system utilises a water/antifreeze coolant to carry away excess energy produced in the form of heat. The cylinder is surrounded by a water jacket from which the heated coolant is circulated by thermo-syphonic action in conjunction with a water pump fitted on the engine left-hand crankcase. On all models the pump is driven from the crankshaft via the clutch and idler drive shaft, but on earlier models the drive is then transmitted via the crankcase breather/water pump drive gear assembly to a driven gear mounted on a separate pump shaft which rotates inside the balancer shaft; on MTX125 RW-H and L models the separate shaft has been deleted and the pump is mounted on the balancer shaft itself. The hot coolant passes upwards through a flexible pipe to the top of the radiator which is mounted on the frame downtubes to take advantage of maximum air flow. The coolant then passes downwards, through the radiator core, where it is cooled by the passing air and then to the water pump and engine where the cycle is repeated.

The complete system is partially sealed and is pressurised; the pressure being controlled by a valve contained in the spring-loaded radiator cap. By pressurising the coolant the boiling point is raised, preventing premature boiling in adverse conditions. The overflow pipe from the radiator is connected to an expansion tank into which excess coolant is discharged by pressure. The expelled coolant automatically returns to the radiator to provide the correct level when the engine cools again.

A thermostat is fitted in the cylinder head cover to increase the speed at which the engine warms up. It shuts off the flow to the radiator until the coolant has warmed up to the point where the thermostat expands to open and permit full flow.

2 Cooling system: draining

1 **Warning:** to avoid the risk of personal injury such as scalding, the cooling system should be drained only when the engine and cooling system is **cold**. Note also that coolant will attack painted surfaces; wash away any spilled coolant immediately with fresh water.

2 On MBX125 models, remove the seat, the right-hand side panel and the left-hand fuel tank panel. On MTX125/200 models remove the radiator (right-hand) shroud which is retained by two screws at the front and one from the side, then remove the expansion tank shroud which is retained by two screws from the side; both shrouds have hooks at the bottom which must be disengaged with care or they may break off.

3 Remove the radiator filler cap. Place a wad of rag over the cap to catch any water or steam ejected under residual pressure, then turn the cap anti-clockwise until a stop is reached. Wait a few moments to allow pressure to equalise (when any hissing noises have stopped) then push the cap downwards and turn it further anti-clockwise to release it.

4 Place the machine upright on a suitable stand and gather together a drain tray or bowl of about 1.5 litre (2.6 pint) capacity, and something to guide the coolant from each drain plug into the bowl. A small chute made from thick card will suffice for this purpose, but do not be tempted to allow the coolant to drain over the engine casings – the antifreeze content may discolour the painted surfaces.

5 The main drain plug is in the bottom of the water pump cover. Position the chute and remove the plug, allowing the coolant to drain fully into the container. A second drain plug is fitted to the front of the cylinder barrel, next to the exhaust port and may be used if required. However it is much easier to tilt the machine over to the left so that as much coolant as possible can drain out. This will be sufficient for all normal draining operations.

6 If the system is to be totally drained, the cylinder head/radiator hose must be disconnected so that the residual coolant left above the thermostat can be removed. Wash away any coolant spilled over the engine and cycle parts as a result of this. The expansion tank should now be emptied.

7 On MTX125/200 models it is sufficient to disconnect the radiator/expansion tank pipe at the radiator union; this can then be directed into the container to empty the tank. On MBX125 models it will be necessary to disconnect and remove the battery and to remove the battery tray and the engine oil tank so that the expansion tank can be removed and emptied.

8 When the coolant is completely drained, refit all disturbed components ensuring that the hose connections are securely fastened. Check that their sealing washers are in good condition and refit the drain plugs, tightening them to a torque setting of 0.8 – 1.2 kgf m (6 – 9 lbf ft). Wash away any spilled coolant.

9 If the coolant is uncontaminated it can be reused, but old or discoloured coolant is best discarded and a new solution used. Remember that the system is prone to corrosion and leakage if the mixture has lost its inhibiting properties. It is recommended that the system is flushed as described below before the new coolant is added.

2.3a MBX125 models – remove fuel tank left-hand panel ...

2.3b ... to gain access to radiator filler cap

2.3c MTX125/200 models – radiator shroud is retained by two screws at front and one from side ...

2.3d ... expansion tank shroud is retained by two screws from the side

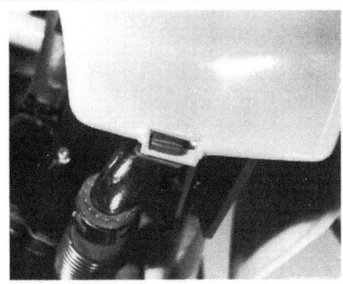

2.3e Be careful not to break bottom mounting hooks when removing shrouds

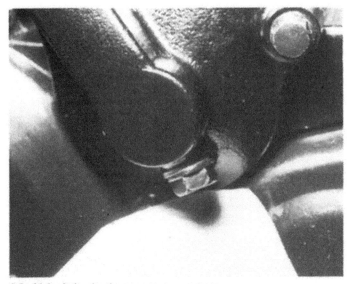

2.5a Main drain plug is on water pump cover ...

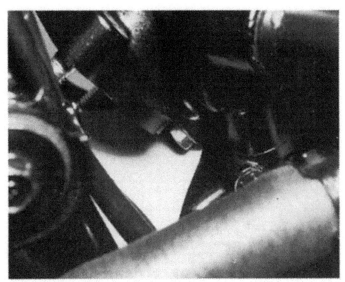

2.5b ... note second drain plug in cylinder barrel – use fabricated chute to keep coolant off crankcases

3 Cooling system: flushing

1   After extended service the cooling system will slowly lose efficiency, due to the build-up of scale, deposits from the water and other foreign matter which will adhere to the internal surfaces of the radiator and water channels. This will be particularly so if distilled water has not been used at all times. Removal of the deposits can be carried out easily, using a suitable flushing agent, in the following manner.

2   After allowing the cooling system to drain, refit the drain plugs and refill the system with clean water and a quantity of flushing agent. Any proprietary flushing agent in either liquid or dry form may be used, providing that it is recommended for use with aluminium engines. NEVER use a compound suitable for iron engines as it will react violently with the aluminium alloy. The manufacturer of the flushing agent will give instructions as to the quantity to be used.

3   Run the engine for ten minutes at operating temperatures and drain

the system. Repeat the procedure TWICE and then again using only clean cold water. Finally, refill the system as described in the following Section.

4 Cooling system: filling

1   The recommended coolant to be used in the system is made up of 30% corrosion-inhibited ethylene-glycol suitable for use in aluminium engines and 70% distilled water; this gives protection against the coolant freezing in temperatures of down to –16°C (3°F). Other mixture ratios of the same ingredients for different temperatures are listed in the Specifications Section of this Chapter to give adequate protection against wind chill factor and other variables, the coolant should always be prepared for temperatures –5° (–9°F) lower than the lowest anticipated.

2   Use only good quality antifreeze of the type specified; never use

alcohol-based antifreeze. In view of the small quantities necessary it is recommended that distilled water is used at all times. Against its extra cost can be set the fact that it will keep the system much cleaner and save the time and effort spent flushing the system that would otherwise be necessary. Tap water that is known to be soft, or rainwater caught in a non-metallic container and filtered before use, may be used in cases of real emergency only. Never use hard tap water; the risk of scale building up is too great.

3   So that a reserve is left for subsequent topping-up, make up approximately 1.5 litres (2.64 pint) of coolant in a clean container. At the **standard** recommended mixture strength this will mean adding 450cc (0.79 pint) of antifreeze to 1050cc (1.85 pint) of distilled water; do not forget to alter the ratio if different temperatures are expected.

4   Having checked the system as described in the subsequent Sections of this Chapter, add the new coolant via the radiator filler neck. Pour the coolant in slowly to reduce the amount of air which will be trapped; when the level is up to the base of the filler neck, fill the expansion tank to its upper level line. This should take approximately 280cc (0.49 pint) on MBX125 models, 150cc (0.26 pint) on MTX125/200 models. Refit the expansion tank filler cap.

5   Start the engine and allow it to idle until it has warmed up to normal operating temperature, with the temperature gauge needle giving its usual reading; the level in the radiator will drop as the coolant is distributed and the trapped air expelled. Add coolant as necessary. As soon as the thermostat opens, revealed by the sudden steady flow of coolant across the radiator and by a warm top hose, the level will drop again and more air will be expelled in the form of bubbles.

6   All trapped air must be expelled from the system before the radiator cap is refitted. When the level has stabilised for some time with the engine fully warmed up, and there are no more signs of air bubbles appearing, top the level up to the base of the filler neck and refit the radiator cap. Stop the engine, check that the expansion tank is topped-up to its upper level mark and refit the radiator shrouds or tank panel, seat and side panel, as appropriate.

7   When the machine has been ridden for the first time after renewing the coolant and has cooled down, check the level again at the radiator cap to ensure that no further pockets of air have been expelled; top up if necessary. At all other times the coolant level should be checked at the expansion tank, as described in Routine Maintenance.

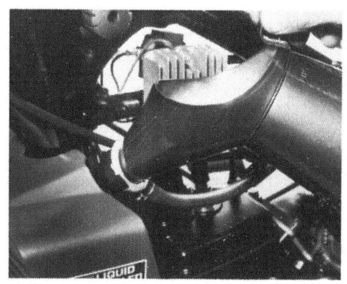

4.4 Mix coolant as described in text and refill via radiator filler

5   Radiator and radiator cap: removal, cleaning, examination and refitting

1   Drain the cooling system as described in Section 2 of this Chapter.

2   On MBX125 models remove its two rear mounting screws and two front mounting nuts and withdraw the belly fairing, then unclip the radiator shroud from its mountings and release it. Disconnect the expansion tank tube at the radiator filler neck. Disconnect the top and bottom hoses as described in Section 6 of this Chapter. Remove the two bottom mounting bolts and lower the radiator away from its top mounting grommet.

3   On MTX125/200 models disconnect the expansion tank tube at the radiator filler neck and secure the tube end above the tank so that no coolant is spilled. Disconnect the top and bottom hoses as described in Section 6 of this Chapter. Remove the single bolt which retains the radiator guard (its other two mounting screws will have been removed with the radiator shroud), withdraw the guard and unscrew the three radiator mounting bolts (two at the front, one at the rear). Withdraw the radiator.

4   On all models, note carefully the relative positions of the mounting rubbers and the metal washers and spacers before removing the mounting bolts. When the radiator is removed, check the mounting rubbers for perishing or compaction. Renew the rubbers if there is any doubt as to their condition. The radiator may suffer from the effects of vibration if the isolating characteristics of the rubber are reduced.

5   The exterior of the radiator core can be cleaned by blowing from the rear surface forwards with compressed air. Most garages have an air line which can be used for this purpose, but if none is available a garden hose can be substituted, using water instead of air, and a fine-bristled toothbrush. It is essential that all dirt and debris are removed from the radiator matrix as any obstructions to the constant

flow of air over its surface will severely reduce the radiator's cooling efficiency. On MTX 125/200 models which are used regularly off-road, great care must be paid to ensuring that the radiator is washed off at every available opportunity so that mud and dirt is not baked on to the matrix; if this is allowed to happen the radiator will be very difficult to clean.

6   The interior of the radiator can most easily be cleaned while the radiator is in place on the motorcycle, using the flushing procedure described in Section 3 of this Chapter. Additional flushing can be carried out by placing the hose in the filler neck and allowing the water to flow through for about ten minutes; in some cases obstructions can be cleared by placing the hose in the bottom hose stub to reverse the flow through the radiator. Under no circumstances should the hose be connected to the filler neck mechanically as any sudden blockage in the radiator outlet would subject the radiator to the full pressure of the mains supply (about 50 psi). The radiator should not be tested to greater than 14 psi (1.05 kg/cm²).

7   If care is exercised, bent fins can be straightened by placing the flat of a screwdriver either side of the fin in question and carefully bending it into its original shape. Badly damaged fins cannot be repaired. If bent or damaged fins obstruct the air flow more than 20%, a new radiator will have to be fitted.

8   Generally, if the radiator is found to be leaking, repair is impracticable and a new component must be fitted. Very small leaks may sometimes be stopped by the addition of a special sealing agent in the coolant. If an agent of this type is used, follow the manufacturer's instructions very carefully. Soldering, using soft solder, may prove effective for caulking large leaks but this is a specialised repair best left to experts.

9   If the radiator cap is suspect, have it tested by a Honda dealer. This job requires specialist equipment and cannot be done at home. The only alternative is to try a new cap, but it should be noted that as the cap is very similar in size to those fitted to cars, a local car garage may have the necessary equipment; therefore, worthwhile to find out if this is the case before going to the possibly unnecessary expense of substituting a new component. If the equipment is available, the cap is fitted to one end, using an adaptor if necessary, and a pressure of 13 psi (0.9 kg/cm²) applied to it, usually by means of a hand-operated plunger. The pressure must be held for a period of 6 seconds, during which time there should be no measurable loss; note that the pressure can, however, vary within a specified tolerance.

10 If the cap is found to be faulty in any way it should be renewed. Note however, that at the time of writing, the radiator cap is not listed as a separate item for these models and can only be purchased (as a Honda replacement part) as part of the complete radiator. The only alternative to this is to approach a good Honda service agent who will

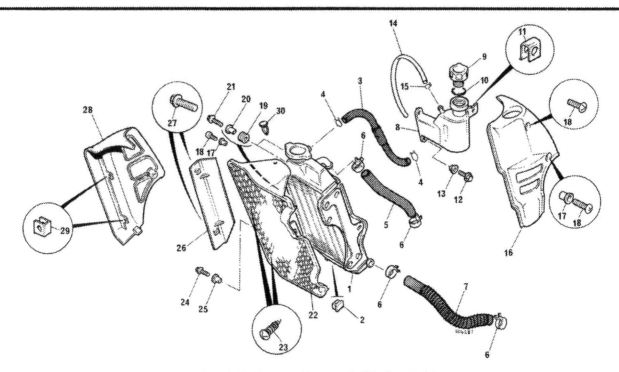

Fig. 2.1 Radiator and hoses – MTX125 and 200

1	Radiator	9	Cap	17	Spacer – 2 off	24	Bolt
2	Rubber	10	O-ring	18	Screw – 3 off	25	Spacer
3	Water hose	11	Nut	19	Grommet – 3 off	26	Plate
4	Clip – 2 off	12	Bolt – 2 off	20	Spacer – 3 off	27	Bolt – 2 off
5	Top hose	13	Spacer – 2 off	21	Bolt – 3 off	28	Shroud
6	Clamp – 4 off	14	Breather tube	22	Grille	29	Nut – 2 off
7	Bottom hose	15	Clip	23	Screw – 3 off	30	Clip
8	Expansion tank	16	Shroud				

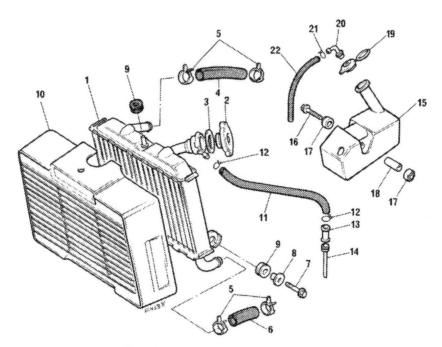

Fig. 2.2 Radiator and hoses – MBX125

1	Radiator	7	Bolt – 2 off	13	Sleeve	18	Spacer
2	Cap	8	Spacer – 2 off	14	Level tube	19	Filler cap
3	Gasket	9	Grommet – 3 off	15	Expansion tank	20	Elbow
4	Top hose	10	Grille	16	Bolt	21	Clip
5	Clamp – 4 off	11	Water hose	17	Damping rubber – 2 off	22	Breather tube
6	Bottom hose	12	Clip – 2 off				

compare the cap with others in his stock (which fit other water cooled
Honda machines) until a suitable cap is found. If this is not possible,
try to find a second-hand but serviceable cap at a motorcycle breakers.
11 It should be noted that when tracing an elusive leak the entire
cooling system can be pressurised to its normal operating pressure by
connecting the test equipment described above to the radiator filler
orifice. Remove the radiator cap, check the coolant level, topping it up
if necessary, and apply a pressure of no more than 15 psi (1.05 kg/cm²)
by means of the hand-operated plunger. Any leaks should soon
become apparent. Most leaks will, however, will be readily apparent
due to the tell-tale traces of antifreeze left on the components in the
immediate area of the leak.
12 On refitting, connect the radiator hoses as described in Section 6 of
this Chapter then tighten the mounting bolts to the specified torque
settings; on MBX125 models, note that the radiator mounting bolts
also retain the belly fairing/shroud mounting brackets. Do not tighten
fully the radiator mounting bolts until the shroud and belly fairing have
been refitted, or there is a risk of distorting and damaging them. Fill the
cooling system as described in Section 4.

5.2a MBX125 models – unhook radiator shroud from top mounting ...

5.2b ... disconnect expansion tank tube at radiator filler neck ...

5.2c ... use pliers as shown to release radiator hose clamps ....

5.2d ... and disconnect radiator hoses ...

5.2e ... unhook radiator from top mounting and withdraw

5.3a MTX125/200 models – single bolt retains radiator guard ...

5.3b ... radiator is held by two bolts at the front ...

5.3c ... and one bolt at the rear

6 Hoses and connections: removal, refitting and checking for leaks

1   The radiator is connected to the engine unit by two flexible hoses, there being an additional pipe between the water pump and the cylinder. The hoses should be inspected periodically and renewed if any sign of cracking or perishing is discovered. The most likely area for this is around the clips which secure each hose to its unions. Particular attention should be given if regular topping up has become necessary. The cooling system can be considered to be a semi-sealed arrangement, the only normal coolant loss being minute amounts through evaporation in the expansion tank. If significant quantities have vanished it must be leaking at some point and the source of the leak should be investigated promptly.

2   To disconnect the hoses, use a large pair of pliers to release the clamps and to slide them along the hose clear of the union spigot. Carefully work the hose off its spigots, noting that it may be necessary to slacken, or remove fully, the radiator mounting bolts to provide room to manoeuvre. The hoses can be worked off with relative ease when

new, or when hot; do not, however attempt to disconnect the system when it is hot as there is a high risk of personal injury through contact with hot components or coolant.

3   **Warning:** the radiator hose unions, and coolant pipes (MBX125 only), are fragile; **do not use excessive force** when attempting to remove the hoses. If a hose proves stubborn, try to release it by rotating it on its unions before attempting to work it off. If all else fails, cut the hose with a sharp knife then slit it at each union so that it can be peeled off in two pieces. While expensive, this is preferable to buying a new radiator.

4   Serious leakage will be self-evident, though slight leakage can be more difficult to spot. It is likely that the leak will only be apparent when the engine is running and the system is under pressure, and even then the rate of escape may be such that the hot coolant evaporates as soon as it reaches the atmosphere, although traces of antifreeze should reveal the source of the leak in most cases. If not, it will be necessary to use testing equipment, as described in the previous Section, to pressurise the cooling system when cold, thereby enabling the source of the leak to be pinpointed. To this end it is best to entrust this work to a Honda service agent who will have access to the necessary equipment if this is not available elsewhere, for example at a car garage or radiator repair agent.

5   In very rare cases the leak may be due to a broken head gasket in which case the coolant may be drawn into the engine and expelled as vapour in the exhaust gases. If this proves to be the case it will be necessary to remove the cylinder head for investigation. If the rate of leakage has been significant it may prove necessary to remove the cylinder barrel and piston so that the crankcase can be checked. Any coolant which finds its way that far into the engine can cause rapid corrosion of the main and big end bearings and must be removed completely.

6   Other possible sources of leakage are the gasket and O-rings sealing the water pump body/crankcase and body/cover joints, the mechanical seal, the seal and O-ring in the pump seal collar, the seals around the spark plug boss and cylinder head cover and on MBX125 models only, the O-rings sealing the coolant metal pipe unions. All these should be investigated and any leaks rectified by tightening the retaining screws, where applicable, or by renewing any seals or gaskets which are worn or damaged.

7   On refitting hoses, first slide the clamps on to the hose and then work it on to each spigot in turn. **Do not** use lubricant of any type; the hose can be softened by soaking it in boiling water before refitting, although care is obviously required to prevent the risk of personal injury when doing this. When the hose is fitted, rotate it to settle it on its spigots and check that the two components being joined are securely fastened so that the hose is correctly fitted before its clamps are slid into position.

6.1 Check all hoses and connections for leaks

wear or damage. Although this is unlikely, check that the impeller blades are not cracked or worn and that the pump body and cover are not cracked; also that there are no marks on any gasket surface. The pump seal collar oil seal and O-ring should be renewed whenever they are disturbed.

4   Check the pump bearing by feeling for free play at the shaft end; the bearing should be removed and refitted as described in Chapter 1. If the drive has been dismantled, check its components for chipped, worn or broken teeth, worn bearing surfaces or other signs of damage; renew any worn or damaged item.

5   Check the mechanical seal for signs of wear or damage, but do not disturb it unless renewal is necessary. The seal itself can be drifted out of the water pump body using as a drift a socket spanner which bears on the seal outer diameter; drive the seal out towards the crankcase side of the body. The seal must be renewed in conjunction with the seal plate, which is removed from the rear of the impeller. If grease is used to assist refitting, ensure that the surplus is wiped away before the body is refitted.

7   Water pump: removal, overhaul and refitting

1   The water pump will not normally require attention unless there is evidence of leakage of coolant into the transmission oil, as shown by excessive amounts of light grey-brown emulsified oil and a persistent drop in the coolant level which cannot be accounted for by any other reason. If a very high mileage has been covered, noise from the water pump area might indicate a worn impeller shaft or bearing.

2   The water pump cover and body, the impeller, the mechanical seal and the pump seal collar can be removed as described in Section 11 of Chaoter 1 and are refitted as described in Section 31. If the pump bearing, shaft or any part of the drive mechanism are to be checked, the engine/gearbox unit must be removed from the frame and completely dismantled until the crankcases can be separated. Refer to Chapter 1.

3   Carefully remove all traces of scale or corrosion from the pump castings and from the impeller, then check carefully for any signs of

7.2 Water pump drive cannot be dismantled without separating crankcases

7.5a Pump mechanical seal should not be disturbed unless it is to be renewed ...

7.5b ... ensure it is correctly located on refitting

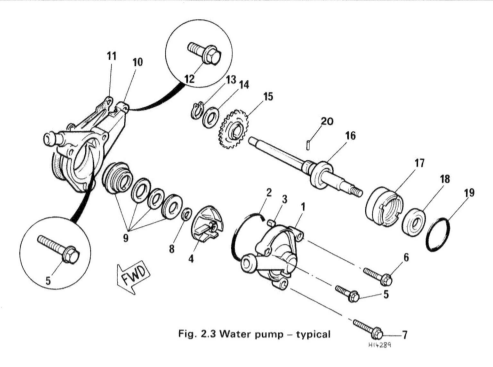

Fig. 2.3 Water pump – typical

1	Pump cover	6	Bolt
2	O-ring	7	Bolt – 3 off
3	Dowel – 2 off	8	Copper washer
4	Impeller	9	Mechanical seal
5	Bolt – 3 off	10	Pump body

11	Gasket	16	Pump drive shaft*
12	Bolt	17	Seal collar
13	Circlip*	18	Oil seal
14	Thrust washer*	19	O-ring
15	Pump driven gear*	20	Drive pin*

*except MTX125 RW-H
model

8 Thermostat: removal and testing

1 The thermostat is housed in the cylinder head cover, which is removed and refitted as described in Section 6 and 36 of Chapter 1.

2 Remove its two mounting bolts from inside the cylinder head cover and withdraw the thermostat and its mounting plate, noting which way round the thermostat is fitted. Tighten the bolts securely on refitting.

3 Thermostat failure is rare, especially if the coolant is renewed at the specified intervals using only distilled water so that corrosion and contamination are kept to a minimum which might cause it to stick. If it does fail, it is most likely to do so in the closed position, thus restricting the flow of coolant and causing severe overheating.

4 To test the unit, remove it from the machine and suspend it in a saucepan of water so that it does not touch the sides or bottom. A thermometer will be required. Start heating the water, noting its temperature and the condition of the thermostat. At 69.5 – 72.5°C (157 – 162.5°F) the thermostat should start to open; it should then open progressively with increasing temperature until 80°C (176°F) is reached, when it will be fully open. Keep the water at least at this temperature for about 5 minutes, then withdraw the thermostat and measure the valve lift; it should be at least 4.5 mm (0.18 in). Allow the thermostat to cool down and check that it closes smoothly.

5 If the thermostat does not open at the specified temperature, if it does not open fully enough, or if it sticks open, it must be renewed; repairs are not possible.

6 As an emergency measure only, if the thermostat is faulty it can be removed and the machine used without it. Take care when starting the engine from cold as the warm-up will take much longer than usual, and ensure that a new unit is fitted as soon as possible.

9 Temperature gauge and sender unit: general

1 An electrically-operated system monitors the engine temperature and thus the cooling system's efficiency. The components, which consist of the sender unit mounted in the cylinder head cover and the gauge mounted in the instrument panel, are tested as described in Chapter 7.

2 To remove the sender unit, disconnect the wire leading to it, then slacken it using a suitable spanner. Unscrew the unit as fast as possible and withdraw it, plugging the opening to stop the coolant escaping. On refitting, coat the unit threads with suitable sealant (Three-Bond No. 1212 or equivalent is recommended) and screw it into place. If sufficient care is taken, the minimum of coolant will escape and there will be no need to drain and refill the system, but remember to top up the radiator or expansion tank, and to wash off all surplus coolant. Tighten the sender unit to a torque setting of 0.8 – 1.4 kgf m (6 – 10 lbf ft) and connect its lead wire again.

8.2a Remove two bolts and mounting plate ...

8.2b ... to remove thermostat from cylinder head cover

9.2 Apply sealant to sender unit threads on refitting – tighten to specified torque setting

Chapter 3 Fuel system and lubrication

Contents

Specifications

Unless otherwise stated, information applies to all models

Fuel tank capacity

	MBX125	MTX125/200
Overall	13.0 litre (2.86 gal)	9.0 litre (1.98 gal)
Including reserve	2.7 litre (0.59 gal)	2.2 litre (0.48 gal) – RW-D model
		2.0 litre (0.43 gal) – RW-F, H model

Fuel grade

Manufacturer's recommendation .... Unleaded or leaded (minimum 91 octane RON)

Carburettor

	MBX125	MTX125	MTX200
Manufacturer	Keihin	Keihin	Keihin
Venturi diameter	24 mm (0.95 in)	24 mm (0.95 in)	26 mm (1.02 in)
ID number	PE60D-A	PE60B-A/C (D model)	PE65A-A (D model)
		PE60B-C/D (F, H models)	PE65A-C (F model)
Main jet	115	112	118
Pilot jet	48	45	48
Needle clip position – grooves from top	3rd	2nd	2nd
Pilot air screw – turns out	$1^1/2$	$1^3/4$	$1^3/4$
Float height	19 mm (0.75 in)	19 mm (0.75 in)	19 mm (0.75 in)
Idle speed	1300 ± 100 rpm	1300 ± 100 rpm	1400 ± 100 rpm

ATAC system – MTX200 only

Drive shaft OD – at left-hand end	9.985 – 10.000 mm (0.3932 – 0.3937 in)
Service limit	9.940 mm (0.3913 in)
Valve closes at	6500 – 6800 rpm

Engine lubrication system

Recommended oil	Good quality two-stroke engine oil suitable for motorcycle injection systems
Tank capacity	1.2 lit (2.1 pint)

Gearbox lubrication

Recommended oil	Good quality SAE10W30 SE or SF engine oil
Capacity:	
At oil change	500 cc (0.9 pint)
At engine rebuild	600 cc (1.1 pint)

Torque wrench settings

Component	kgf m	lbf ft
Fuel tap filter bowl	0.3 – 0.5	2 – 3.5
Air filter case mounting bolt	0.9 – 1.3	6.5 – 9.5
Inlet stub mounting bolts	0.8 – 1.2	6 – 9
ATAC valve assembly and operating spindle mountings – MTX200	1.8 – 2.0	13 – 14.5
Exhaust pipe/cylinder barrel retaining nuts	1.8 – 2.5	13 – 18
Exhaust pipe rear mounting – MBX125	2.0 – 2.4	14.5 – 17
Exhaust pipe and silencer mounting bolts – MTX125/200	1.8 – 2.5	13 – 18
Oil pump mounting bolt	0.8 – 1.2	6 – 9
Oil tank mounting bolt:		
MBX125	0.9 – 1.3	6.5 – 9.5
MTX125/200	0.8 – 1.2	6 – 9
Transmission oil drain plug	2.0 – 2.5	14.5 – 18

1 General description

The fuel system comprises a petrol tank, from which petrol is fed by gravity to the float chamber of the carburettor, via a three position petrol tap. The tap has 'Off', 'On' and 'Reserve' positions, the latter providing a warning that the petrol level is low in time for the owner to find a garage.

The carburettor is of conventional concentric design, the float chamber being integral with the lower part of the carburettor. Cold starting is assisted by a separate starting circuit which supplies the correct fuel-rich mixture when the 'choke' control is operated.

Air entering the carburettor passes through an air cleaner casing which contains an oil-impregnated foam air filter. This effectively removes any airborne dust, which would otherwise enter the engine and cause premature wear. The air cleaner also helps silence induction noise, a common problem with two-stroke engines.

A reed valve is fitted in the intake tract to help the engine's performance at low to medium speeds. Consisting of two reed 'petals' mounted on an alloy valve body, the petals open and close automatically as a result of the difference between atmospheric pressure and the pressure in the crankcase as the engine is running. Their movement is controlled by two carefully-shaped stopper plates.

On MTX200 models only the ATAC (Automatically-controlled Torque Amplification Chamber) system is fitted to the exhaust to boost the power output at low to medium engine speeds. Since any two-stroke exhaust system can be tuned only to produce maximum power at one engine speed, at the expense of all other engine speeds, if the exhaust's volume, and therefore its effective length, can be altered, the performance range will be much improved. In the Honda system a HERP (Honda Energy Resonance Pipe) chamber is provided in the form of passages cast into the cylinder head and barrel castings; this is open to the exhaust at low to medium engine speeds and is closed off by a valve at higher speeds. The valve is controlled automatically by a centrifugally-operated mechanical governor which rotates a spindle via a rack; the spindle is linked to the valve outside the crankcases.

Engine lubrication is catered for by a pump-fed total loss system. Oil from a separate tank is fed by an oil pump to a small injection nozzle in the intake tract. The pump is linked to the throttle twistgrip, and this controls the volume of oil fed to the engine.

All gearbox and primary drive components are lubricated by splash from a supply of oil contained in the reservoir formed by the crankcase castings.

2 Fuel tank: removal, examination and refitting

1 On MBX125 models, unlock and remove the seat; then carefully remove both side panels. Switch the fuel tap to the 'Off' position and use a pair of pliers to disengage the pipe retaining clip and to slide it down the pipe clear of the tap spigot; pull off the feed pipe. Remove the tank rear mounting bolt, lift the tank at the rear and pull it backwards off its front mounting rubbers.

2 On MTX125/200 models unbolt and remove the seat, then

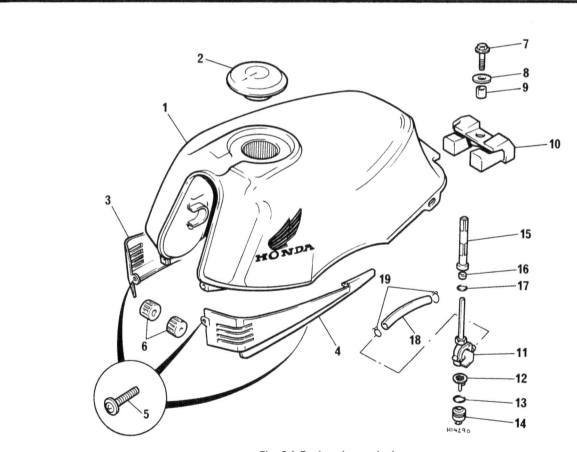

Fig. 3.1 Fuel tank – typical

1 Fuel tank	6 Mounting rubber – 2 off	11 Fuel tap	16 Collar
2 Filler cap	7 Bolt	12 Filter gauze	17 O-ring
3 Right-hand panel	8 Washer	13 O-ring	18 Fuel pipe
4 Left-hand panel	9 Spacer	14 Filter bowl	19 Clip – 2 off
5 Screw and washer – 4 off	10 Mounting rubber	15 Filter gauze	

carefully remove both side panels. Remove the radiator shrouds as described in Section 2 of Chapter 2 and release the filler cap breather tube. Remove the fuel tank, disconnecting first the feed pipe, as described above.

3   On refitting, check that all mounting rubbers are in good condition, renewing any that are perished, split, compacted or otherwise damaged. If they are tight, lubricate the front mounting rubbers with soapy water or similar and wiggle the tank from side to side to slide it fully into place. Tighten the mounting bolt securely, ensuring that the metal spacer is correctly refitted, and check carefully for fuel leaks after the feed pipe has been refitted and secured and the tap has been switched on. Check that there is no metal-to-metal contact between tank and frame and that the tank is completely isolated from vibration by its mounting rubbers.

4   Store the tank in a safe place whilst it is removed from the machine, well away from any naked lights or flames. It will otherwise represent a considerable fire or explosion hazard. Check that the tap is not leaking and that it cannot be accidentally knocked into the 'On' position. Placing the tank on a soft protected surface. Covering it with a protective cloth or mat may well avoid damage being caused to the finish by dirt, grit, dropped tools, etc.

5   If fuel tank repairs are necessary the following points should be noted. Fuel tank repair, whether necessitated by accident damage or by fuel leaks, is a task for the professional. Welding or brazing is not recommended unless the tank is purged of all petrol vapour; which is a difficult condition to achieve. Resin-based tank sealing compounds are a much more satisfactory method of curing leaks, and are now available through suppliers who advertise regularly in the motorcycle press. Accident damage repairs will inevitably involve re-painting the tank; matching of modern paint finishes, especially metallic ones, is a very difficult task to be lightly undertaken by the average owner. It is therefore recommended that the tank be removed by the owner, and then taken to a motorcycle dealer or similar expert for professional attention.

6   Repeated contamination of the fuel tap filter and carburettor by water or rust and paint flakes indicates that the tank should be removed for flushing with clean fuel and internal inspection. Rust problems can be cured by using a resin tank sealant.

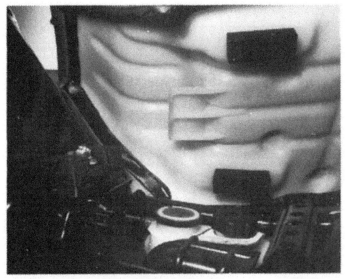
2.1a Remove seat to expose tank rear mounting

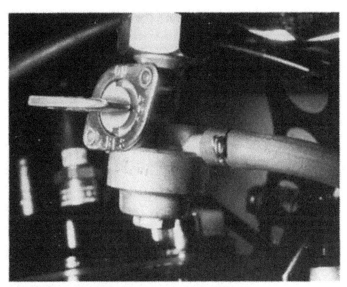
2.1b Switch off tap and disconnect fuel feed pipe

2.1c Remove tank rear mounting bolt to release tank

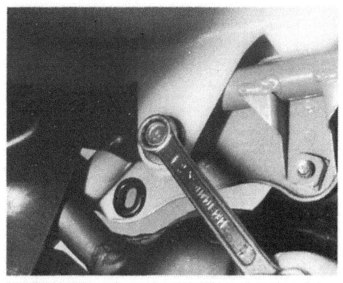
2.2 MTX125/200 models – seat is unbolted from underneath one bolt each side

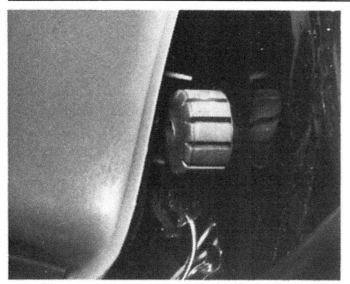

2.3 Ensure tank guide channels engage fully on front mounting rubbers

4.2 Slacken retaining clamps and release breather and drain tubes to withdraw carburettor

3 Fuel tap and feed pipe: general

Refer to Routine Maintenance for details of all maintenance operations on these components. If the tap is to be removed, the tank should be drained before removal, or if the fuel level is low enough, it should be removed and laid on its side so that petrol cannot escape either from the filler cap or from the tap orifice.

4 Carburettor: removal and refitting

1   As a general rule, the carburettor should be left alone unless it is in obvious need of overhaul. Before a decision is made to remove and dismantle, ensure that all other possible sources of trouble have been eliminated. This includes the more obvious candidates such as a fouled spark plug, a dirty air filter element or choked exhaust system. If a fault has been traced back to the carburettor, proceed as follows.

2   Switch off the fuel tap and disconnect the feed pipe at the tap spigot, as described in Section 2. Slacken fully the clamps securing the air filter hose and inlet stub, release the breather and float bowl drain tubes, then carefully twist the carburettor clear of its mountings and unscrew the top. Withdraw the throttle valve (slide) assembly. If the carburettor is being removed to trace a fault, unscrew the float bowl drain plug and empty the remaining fuel into a clean container; any signs of dirt or water should be obvious and will require further investigation.

3   To disconnect the throttle valve from the cable, hold the carburettor top, compress the throttle return spring against it and hold it. Invert the throttle valve and shake out the spring seat. This component serves to prevent the cable from becoming detached when in position and once out of the way the cable can be pushed down and slid out of its locating groove. The retaining spring and needle can now be removed from the valve.

4   The carburettor is refitted by reversing the removal sequence. Note that it is important that the instrument is mounted vertically to ensure that the fuel level in the float bowl is correct. A locating tab is fitted to provide a good guide to alignment but it is worthwhile checking this for accuracy. Once refitted, check the carburettor adjustments as described later in this Chapter. Note too that the oil pump cable adjustment must be checked after the throttle cable has been adjusted. Route the carburettor breather pipe in underneath the oil pump cover and the float bowl drain tube down the back of the crankcase, securing it with the clamps provided.

5   **Note:** If the carburettor is to be set up from scratch it is important to check jet and float level settings prior to installation. Refer to Sections 6 and 7 before the carburettor is refitted.

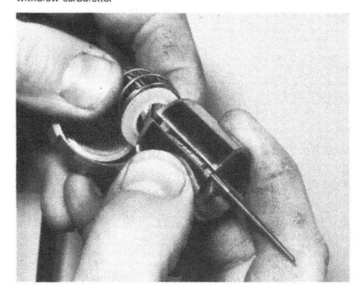

4.3a Removing throttle valve assembly from cable

4.3b Note how retaining clip is fitted before removing needle from throttle valve

4.4 Ensure throttle valve is correct way round on refitting – align valve groove with locating pin (arrowed)

5 Carburettor: dismantling, examination and reassembly

1 Detach the float chamber by inverting the carburettor and removing the retaining screws. There is a sealing O-ring around the edge of the float chamber, which will remain in the float chamber.
2 Pull out the pivot pin from the twin float assembly. Lift away the floats and the float needle together, being careful that the small float needle does not drop free and become lost.
3 The main jet is located in the centre of the mixing chamber, between the twin floats. It is threaded into the base of the needle jet holder (in the central column). When unscrewing any jet, a close fitting screwdriver must be used to prevent damage to the slot in the soft jet material. On MTX125/200 models, remove the white nylon anti-surge baffle from around the jet. On all models, unscrew and remove the main jet. Using a small close fitting spanner, unscrew and remove the needle jet holder from below the main jet (when inverted).
4 Return the carburettor to its upright position and allow the needle jet to fall out. It may be necessary to use a hammer and a soft wooden probe to push the jet out from above.
5 Unscrew the slow (pilot) jet from its location in front of the central column. If the removal of the pilot air screw is necessary, screw it in until it seats lightly and count and record the number of turns necessary to achieve this so that the screw can be returned to its original position on refitting. Unscrew the air screw and remove it together with its spring; the throttle top screw is removed in the same way, if required.
6 After an extended period of service, the throttle slide will wear and may produce a clicking sound within the carburettor body. Wear will be evident from inspection, usually at the base of the slide and in the locating groove. A worn slide should be removed as soon as possible because it will give rise to air leaks which will upset carburation. Check that the needle is not bent by rolling it on a flat surface such as a sheet of plate glass. If it is bent it must be renewed. Always renew the needle and needle jet together.
7 Inspect the choke assembly. Unscrew the operating arm pivot screw and withdraw the arm; renew it if it is cracked, broken or distorted. Unscrew the large retaining nut and withdraw the plunger assembly. If the plunger does not operate smoothly and easily, if the spring or any other component is broken, or if the plunger itself is worn, the assembly must be renewed. Check that the plunger seating is clean and unworn and that the related passages and drillings in the carburettor body are clear.

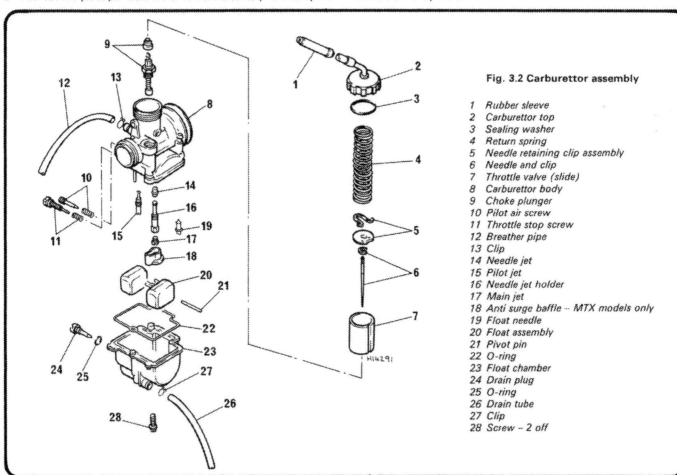

Fig. 3.2 Carburettor assembly

1   Rubber sleeve
2   Carburettor top
3   Sealing washer
4   Return spring
5   Needle retaining clip assembly
6   Needle and clip
7   Throttle valve (slide)
8   Carburettor body
9   Choke plunger
10 Pilot air screw
11 Throttle stop screw
12 Breather pipe
13 Clip
14 Needle jet
15 Pilot jet
16 Needle jet holder
17 Main jet
18 Anti surge baffle – MTX models only
19 Float needle
20 Float assembly
21 Pivot pin
22 O-ring
23 Float chamber
24 Drain plug
25 O-ring
26 Drain tube
27 Clip
28 Screw – 2 off

8 Check carefully the carburettor body, the float chamber, and the carburettor top, looking for signs of damage such as cracks or splits, worn threads, and distorted mating surfaces. Whilst it may be possible in some cases to repair such damage, for example by rubbing flat a distorted mating surface with a sheet of plate glass and fine emery paper, it will be necessary usually to renew the damaged component.

9 Check that the floats are in good order and not punctured. If either float is punctured, it will produce the wrong petrol level in the float chamber, leading to flooding and an over-rich mixture. Because the floats are moulded from a plastic material, it is not possible to effect a permanent repair. In consequence, a new replacement must be fitted if damage is found. Flooding may be caused by an incorrect float height, rather than a damaged float. If this is the case refer to Section 7 for the adjustment procedure.

10 The float needle and needle seat will wear after lengthy service and should be closely examined with a magnifying glass. Wear usually takes the form of a ridge or groove, which will cause the float needle to seat imperfectly. If damage to the seat has occurred, the carburettor body must be renewed because the seat is not removable.

11 Before the carburettor is reassembled by reversing the dismantling procedure, it should be cleaned out thoroughly using compressed air. Avoid using a piece of rag since there is always risk of particles of lint obstructing the internal passageways or the jet orifices.

12 Never use a piece of wire or any pointed metal object to clear a blocked jet. It is only too easy to enlarge the jet under these circumstances and increase the rate of petrol consumption. If compressed air is not available, a blast of air from a tyre pump will usually suffice. As a last resort, a fine nylon bristle may be used.

13 Do not use excessive force when reassembling a carburettor because it is too easy to shear a jet or some of the smaller screws. Furthermore, the carburettor is cast in a zinc-based alloy which itself does not have a high tensile strength. If any of the castings are damaged during reassembly, they will almost certainly have to be renewed.

5.1 Remove retaining screws and withdraw float chamber – renew O-ring if damaged

5.2a Pull out float pivot pin ...

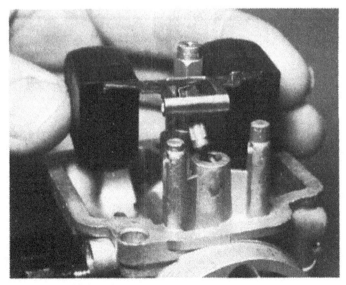

5.2b ... and withdraw float assembly – do not lose float needle

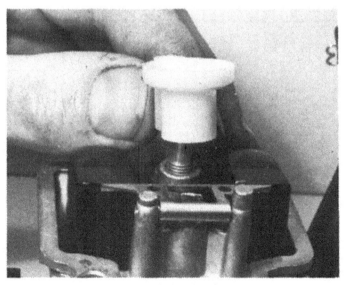

5.3a Remove anti-surge baffle – MTX125/200 models only

5.3b Use close-fitting screwdriver to unscrew main jet ...

5.3c ... then unscrew needle jet holder

5.5a Unscrew pilot jet – check that bleed holes are clear

5.5b If necessary, remove pilot air screw as described in text

5.7a Remove choke lever pivot screw and withdraw lever ...

5.7b ... then unscrew choke plunger assembly

6 Carburettor: checking the settings

1   The various jet sizes, throttle valve cutaway (on slide type carburettors) and needle position are predetermined by the manufacturer and should not require modification. Check with the Specifications list at the beginning of this Chapter is there is any doubt about the types fitted. If a change appears necessary it can often be attributed to a developing engine fault .unconnected with the carburettor. Although carburettors do not wear in service, this process occurs slowly over an extended length of time and hence wear of the carburettor is unlikely to cause sudden or extreme malfunction. If a fault does occur, check first other main systems, in which a fault may give similar symptoms, before proceeding with carburettor examination or modification.

2   Where non-standard items, such as exhaust systems or air filters have been fitted to a machine, some alterations to carburation may be required. Arriving at the correct settings often requires trial and error, a method which demands skill borne of previous experience. In many cases the manufacturer of the non-standard equipment will be able to advise on correct carburation changes.

3   As a rough guide, up to $1/8$ throttle is controlled by the pilot jet, $1/8$ to $1/4$ by the throttle valve cutaway, $1/4$ to $3/4$ throttle by the needle position and from $3/4$ to full by the size of the main jet. These are only approximate divisions, which are by no means clear cut. There is a certain amount of overlap between the various stages.

4   If alterations to the carburation must be made, always err on the side of a slightly rich mixture. A weak mixture will cause the engine to overheat which, particularly on two-stroke engines, may cause engine seizure. Reference to Routine Maintenance will show how, after some experience has been gained, the condition of the spark plug electrodes can be interpreted as a reliable guide to mixture strength.

7 Carburettor: adjustment

1   Before any dismantling or adjustment is undertaken, eliminate all other possible causes of running problems, checking in particular the spark plug, ignition timing, air cleaner and the exhaust. Checking and cleaning these items as appropriate will often resolve a mysterious flat spot or misfire.

2   The first step in carburettor adjustment is to ensure that the jet sizes, needle position and float height are correct, which will require the removal and dismantling of the carburettor as described in Sections 4 and 5 of this Chapter.

3   If the carburettor has been removed for the purpose of checking jet sizes, the float level should be measured at the same time. It is unlikely that once this is set up correctly there will be a significant amount of variation, unless the float needle or seat have worn. These should be checked and renewed, if necessary, as described in Section 5.

4   Remove the float bowl from the carburettor body, if this has not already been done. Check that the gasket surface of the carburettor body is clean and smooth once the gasket is removed. Hold the carburettor body so that the venturi is now vertical with the air filter side upwards and the floats hanging from their pivot pin. Carefully tilt the carburettor to an angle of about 60 – 70° from the vertical so that the tang of the float pivot is resting firmly on the float needle and the float valve is therefore closed, but also so that the spring-loaded pin set in the float needle itself is not compressed. Measure the distance between the gasket face and the bottom of one float, in line with the main jet, with an accurate ruler or a vernier caliper; the distance should be 19 mm (0.75 in).

5   If adjustment is required, remove the float assembly and bend by a very small amount the small tang which acts on the float needle pin. Reassemble the float and measure the height again. Repeat the process until the measurement is correct, then check that the other float is at exactly the same height as the first. Bend the pivot very carefully and gently if any difference is found between the heights of the two floats.

6   When the jet sizes have been checked and reset as necessary, reassemble the carburettor and refit it to the machine as described in Sections 4 and 5 of this Chapter.

7   Start the engine and allow it to warm up to normal operating temperature, preferably by taking the machine on a short journey. Stop the engine and screw the pilot air screw in until it seats lightly, then unscrew it by the number of turns shown in the Specifications Section for the particular model. Start the engine and set the machine to its specified idle speed by rotating the throttle stop screw as necessary. Try turning the pilot air screw inwards by about $1/4$ turn at a time, noting its effect on the idling speed, then repeat the process, this time turning the screw outwards.

8   The pilot air screw should be set in the position which gives the fastest consistent tickover. The tickover speed may be reduced further, if necessary, by unscrewing the throttle stop screw the required amount. Check that the engine does not falter and stop after the throttle twistgrip has been opened and closed a few times.

9   Throttle cable adjustment should be checked at regular intervals and after any work is done to the carburettor, oil pump or to the cable itself. Refer to Routine Maintenance, and do not forget to adjust the oil pump cable after the throttle cable is adjusted.

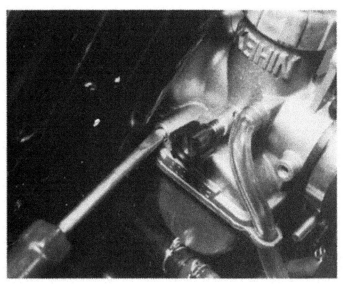

7.7 Adjusting pilot air screw

7.8 Set idle speed using throttle stop screw

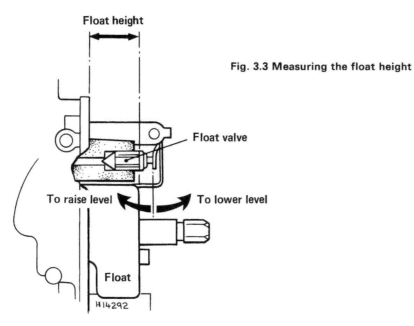

Fig. 3.3 Measuring the float height

8   Air filter: general

The care and maintenance of the air filter element is described in Routine Maintenance. Never run the engine with the air filter disconnected or the element removed. Apart from the risk of increased engine wear due to unfiltered air being allowed to enter, the carburettor is jetted to compensate for the presence of the filter and a dangerously weak mixture will result if the filter is omitted. If the air filter/carburettor hose or air filter intake duct are separated from the casing at any time, they should be glued in place on refitting to ensure an air-tight joint; Honda recommend the use of Cemedine No 540.

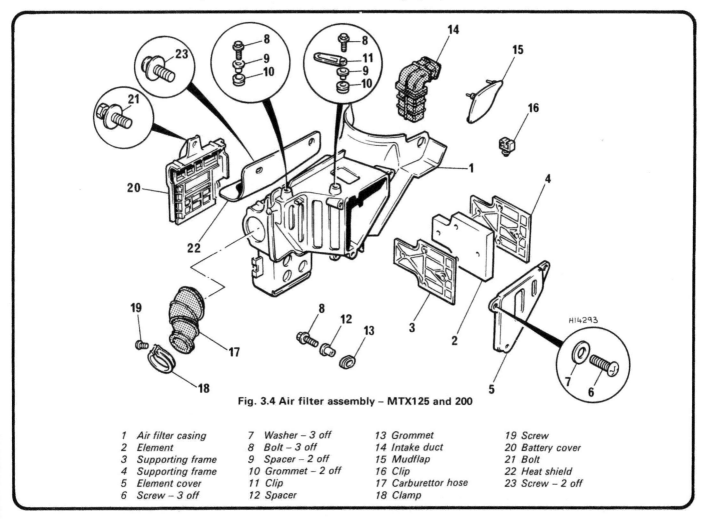

Fig. 3.4 Air filter assembly – MTX125 and 200

1 Air filter casing	7 Washer – 3 off	13 Grommet
2 Element	8 Bolt – 3 off	14 Intake duct
3 Supporting frame	9 Spacer – 2 off	15 Mudflap
4 Supporting frame	10 Grommet – 2 off	16 Clip
5 Element cover	11 Clip	17 Carburettor hose
6 Screw – 3 off	12 Spacer	18 Clamp

19 Screw
20 Battery cover
21 Bolt
22 Heat shield
23 Screw – 2 off

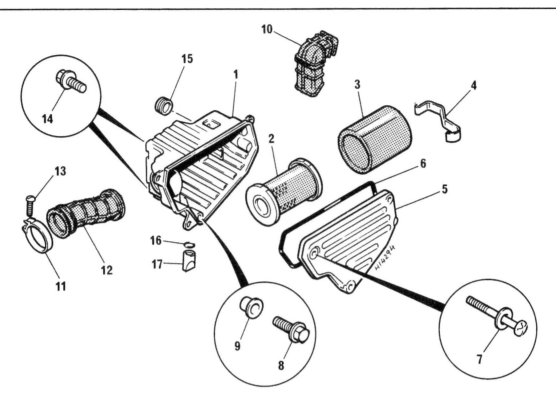

Fig. 3.5 Air filter assembly – MBX125

1	Air filter casing	6	O-ring	10	Intake duct	14	Screw – 2 off
2	Element frame	7	Screw – 3 off	11	Clamp	15	Grommet
3	Element	8	Bolt	12	Carburettor hose	16	Clip
4	Retaining spring clip	9	Spacer	13	Screw	17	Drain tube
5	Element cover						

9 Reed valve assembly: removal, examination and refitting

1   Remove the oil feed pipe from the inlet stub union and plug it to prevent the loss of oil or the entry of air or dirt. Remove the carburettor as described in Section 4 of this Chapter. Slacken and remove the four inlet stub retaining bolts, and displace the stub. It may need a gentle tap using a soft-faced mallet to free it. Remove the reed valve assembly. Take great care when handling the reed valve assembly as it is easily damaged.

2   The reed valve assembly provides a supplementary method of controlling the inlet timing which functions in addition to the normal piston porting arrangement. The normal piston-ported intake tract is designed to open earlier and close later than is usual in engines of this type, thus producing much better high-speed performance at the expense of the low to medium speed ranges. The reed valve operates automatically, opening and closing as a result of the combination of atmospheric pressure and piston position, which in practice has the effect of making the engine more efficient at the low to medium speed ranges. Piston porting and reed valve therefore combine to give the engine more power at all engine speeds than can be available to a conventionally-ported unit. The reed valve has an added advantage in that it eliminates the blow-back of air-fuel mixture through the carburettor which is an inevitable product of a piston-ported engine, thus reducing petrol consumption.

3   The reed valve requires no maintenance as it extremely simple in construction and is automatic in operation. However it is also extremely delicate and must be kept clean at all times and handled very carefully. Check the whole assembly for cracks or other signs of wear, and make sure that the reed petals seat firmly on the rubber valve seat. Any sign of damage at all will mean the whole assembly will have to be renewed. No repairs are possible, and it is not advisable to attempt to strip the assembly further as no individual components are available. Furthermore do not attempt to modify the assembly by bending the stopper plate or by any other means as this will at least adversely affect engine performance. More probably the stress limits will cause the

reeds to crack, allowing the pieces to drop straight into the engine with disastrous consequences.

4   Refitting the reed valve assembly is a straightforward reversal of the removal procedure. Always fit new gaskets to prevent induction leaks and be careful not to overtighten the inlet stub bolts; note, and use if possible, the specified torque setting. When connecting the oil feed pipe, if the end of the pipe is full of oil when it is unplugged, bleeding air from the system will not be necessary. Simply push the end of the pipe over the inlet stub union and secure it with the wire clip. If however, no oil is visible, the feed pipe will have to be bled as described in Section 15 of this Chapter.

10 Exhaust system: general

1   The exhaust is removed and refitted as described in Sections 4 and 37 of Chapter 1. Always renew the exhaust port gasket to prevent leaks, and ensure that all fasteners are tightened to the specified torque settings.

2   The exhaust must be removed at regular intervals for the carbon deposits to be cleaned out. Refer to Routine Maintenance.

3   Do not modify the exhaust system in any way. It is designed to give the maximum power possible consistent with legal requirements and yet to produce the minimum noise level possible. Quite apart from the legal aspects as applied to restricted machines, it is very unlikely that an unskilled person could improve the performance of any of the machines by working on the exhaust. If an aftermarket accessory system is being considered, check very carefully that it will maintain or increase performance when compared with the standard system. Very few 'performance' exhaust systems live up to the claims made by the manufacturers, and even fewer offer any performance increase at all over the standard component.

4   The final point to be borne in mind is the finish. A matt black-painted finish is cheaper to renovate but is less durable than a conventional chrome-plated system. It is inevitable that the original

finish will deteriorate to the point where the system must be removed from the machine and repainted, therefore some thought must be given to the type of paint to be used. Several alternative finishes are offered. Reference to the advertisements in the national motorcycle press, or to a local Honda dealer and to the owners of machines with similarly-finished exhausts will help in selecting the most effective finish. The best are those which require the paint to be baked on, although some

aerosol sprays are almost as effective. Whichever finish is decided upon, ensure that the surface is properly prepared according to the paint manufacturer's instructions and that the paint itself is correctly applied.

5 If any part of the exhaust system is cracked, split or holed through damage or corrosion, it must be renewed immediately to prevent excessive noise and a reduction in performance.

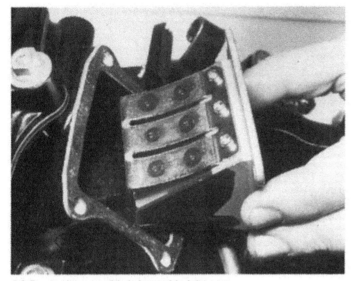
9.2 Reed valve assembly is located in inlet tract

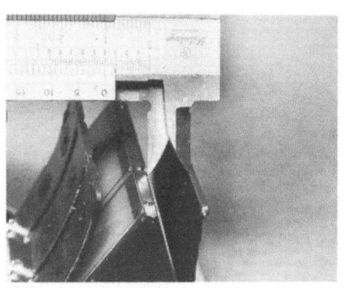
9.3 Measuring stopper plate height – do not bend stopper plates

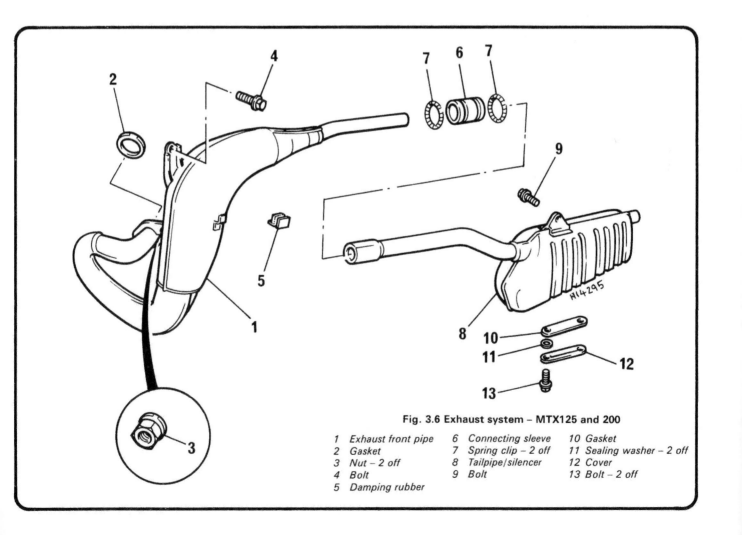

Fig. 3.6 Exhaust system – MTX125 and 200

1 Exhaust front pipe	6 Connecting sleeve	10 Gasket
2 Gasket	7 Spring clip – 2 off	11 Sealing washer – 2 off
3 Nut – 2 off	8 Tailpipe/silencer	12 Cover
4 Bolt	9 Bolt	13 Bolt – 2 off
5 Damping rubber		

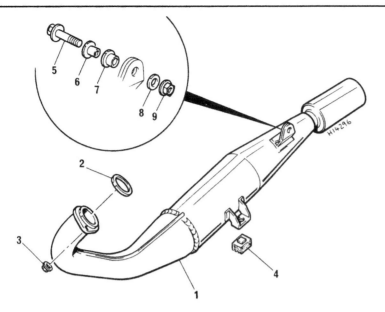

**Fig. 3.7 Exhaust system –
MBX125**

1 Exhaust pipe
2 Gasket
3 Nut – 2 off
4 Stand rubber
5 Bolt
6 Spacer
7 Mounting rubber
8 Washer
9 Nut

11 ATAC system: checking – MTX200

1 In practice it will probably prove difficult to tell, just by riding the machine, whether the ATAC system is functioning or not. The system can only be checked as follows whenever any part of the system has been disturbed or if a loss of power is being investigated.
2 Slacken its clamp and peel back the rubber cover. Start the engine, allow it to warm up and increase engine speed to 6500 – 6800 rpm;

from its fully open position (ie, across to the right) the valve should move smoothly and steadily to its fully closed position; without distorting or straining the connecting arm, press lightly on the valve stem to check that it is fully seated. Slowly close the throttle and check that the valve returns smoothly to the open position. Stop the engine and refit the rubber cover, securing its clamp.
3 If the valve sticks in the open or closed position or if its movement is stiff and jerky and if it does not seat fully, the system must be dismantled for cleaning and checking. Refer to the next Section.

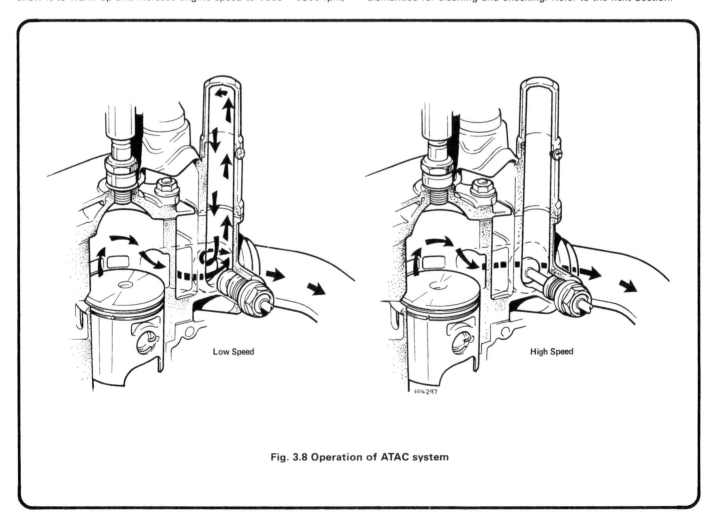

Low Speed   High Speed

Fig. 3.8 Operation of ATAC system

12 ATAC system: removal, examination and refitting – MTX200

1   **Note:** While this system is not, at the time of writing, fitted to any 125cc models sold in the UK, the HERP chamber passages are cast in to the cylinder head and barrel passages and should be checked regularly as described in paragraph 9 below.

2   The system requires no regular maintenance other than to remove and clean the valve, as described below, at the regular intervals given in Routine Maintenance. The cap(s) sealing the HERP chamber passage in the cylinder head should not be disturbed unless necessary; ensure that a leakproof fit is obtained on refitting and that the retaining screw(s) are securely fastened. Apply a few drops of oil to the linking arm whenever the cover is removed.

3   Refer to the following Sections of Chapter 1 for full information on removal and refitting; the disconnecting of the spindle from the valve assembly is described in Section 6 and re-connecting in Section 36. The removal and refitting of the operating spindle is described in Sections 7 and 35 respectively, the removal and refitting of the shaft in Sections 9 and 33. All components should be lubricated on refitting; see below.

4   To remove the valve assembly, disconnect the connecting arm and unscrew the valve mounting nut from the cylinder barrel then withdraw the assembly as a single unit. On refitting, ensure that all components are greased and that the sealing washer is renewed, then insert the assembly into the cylinder barrel and tighten the mounting nut to a torque setting of 1.8 – 2.0 kgf m (13 – 14.5 lbf ft). Connect the valve assembly to the operating spindle and check the system is operating correctly, as described in the previous Section of this Chapter.

5   If the system fails in any way, the reason is most likely to be carbon deposits on the valve stem causing the valve to stick, or corrosion on the operating spindle and/or connecting arm, which will have a similar effect. In this case the various components must be removed and dismantled for thorough cleaning. If the check described in the previous Section shows the mechanical governor to be at fault (ie the system operating at the wrong engine speed, operating jerkily, or not moving far enough so that the valve does not seat properly) the shaft must be renewed; it is carefully set at the factory and should not be disturbed. Since the shaft is fully enclosed and lubricated, wear and corrosion should not be a problem, although it is worth cleaning the shaft in solvent to check that it has not become jammed by dirt, swarf or other foreign matter before renewing it. If the shaft left-hand bearing surface has worn to the service limit specified, or if there are any other signs of wear or damage, the shaft should be renewed immediately; repairs are not possible.

6   To check the operating spindle, pull it out of the mounting nut (once the assembly has been removed) and check that the pulley-type wheel is unworn and free to rotate, also that there is no wear between the spindle and nut; refit the two and feel for free play. The spindle must rotate smoothly in the nut; renew either or both components as necessary if wear is found. Renew the oil seal and sealing O-ring whenever they are disturbed. The components of the connecting arm should be carefully cleaned and checked for damage or for wear at all points of contact; renew any worn or damaged item.

7   Once the valve assembly has been removed, hold the valve firmly but without damaging it and unscrew the retaining nut. Withdraw the plain washer and the spacer, then tap away the retainer and withdraw the two split collets. Remove the valve from the mounting nut and carefully prise out the oil seal. On refitting, press a new oil seal into the mounting nut and smear its sealing lips and the valve stem with silicone grease before refitting the valve. Place the split collets on the valve stem, using silicone grease to retain them and ensuring that they are securely located, then slide the retainer down the valve stem and over the collets. Tap the retainer gently into place then refit the spacer, the plain washer and the retaining nut. Tighten the nut securely and fit a new sealing washer to the assembly mounting nut before refitting it to the machine.

8   The valve assembly should be carefully cleaned, after removal and

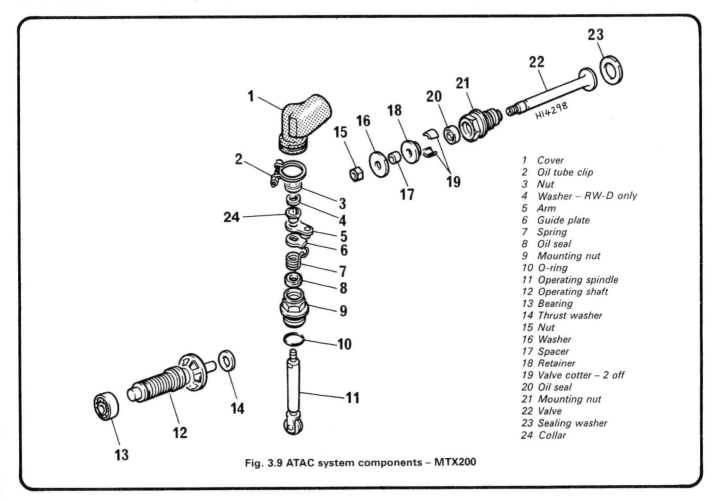

1	Cover
2	Oil tube clip
3	Nut
4	Washer – RW-D only
5	Arm
6	Guide plate
7	Spring
8	Oil seal
9	Mounting nut
10	O-ring
11	Operating spindle
12	Operating shaft
13	Bearing
14	Thrust washer
15	Nut
16	Washer
17	Spacer
18	Retainer
19	Valve cotter – 2 off
20	Oil seal
21	Mounting nut
22	Valve
23	Sealing washer
24	Collar

Fig. 3.9 ATAC system components – MTX200

12.2 Check that HERP chamber cap fits securely with no leaks – all models

12.9 Do not forget to clean HERP chamber passages when decoking engine

dismantling, to remove all carbon deposits from the valve and the mounting nut. Use a soft scraper to remove all large deposits, followed by careful polishing with fine emery paper or similar; do not forget to check the bore of the mounting nut. Check that the valve slides smoothly and easily in the mounting nut and renew any component that is found to be in any way worn or damaged.

9 The cylinder barrel and head HERP chamber passages should be cleaned whenever the top end is dismantled to prevent the excessive build-up of carbon; use a suitable scraper to remove the bulk of the deposits and finish off with wire wool or abrasive paper. **Do not** remove the cap sealing the cylinder head passage unless absolutely necessary; on MTX200 models the cap sealing gasket should be renewed whenever it is disturbed. Tighten the retaining screws securely. On 125 models, check that there are no leaks at the cap joint; a suitable sealant can be used if necessary.

10 On MTX200 models, check that the valve seat in the cylinder barrel is clean and unmarked. While this seating is not as critical as those on a four-stroke cylinder head, the valve should be able to seal off the HERP

chamber with reasonable efficiency. If the seating is scratched or otherwise damaged use **only** fine valve-grinding compound to lap the valve **lightly** on to its seat; do not remove excessive material or the cylinder barrel must be renewed. As soon as a smooth and unbroken seating surface is obtained, stop the operation and wash away all traces of grinding compound. Rebuild temporarily the valve assembly, fit it to the cylinder barrel and push the valve fully across to ensure that it can seat properly; if this is not possible the valve, or even the cylinder barrel, must be renewed.

11 Be careful to flush out the HERP chamber passages with a suitable solvent so that all particles of carbon and cleaning material are removed before the cylinder head and barrel are refitted.

12 On refitting, the valve stem must be smeared with silicone grease as described above. The shaft and operating spindle should be smeared with molybdenum disulphide-based grease, as should the threads of the spindle and valve mounting nuts. Apply a few drops of oil to the connecting shaft before refitting its rubber cover.

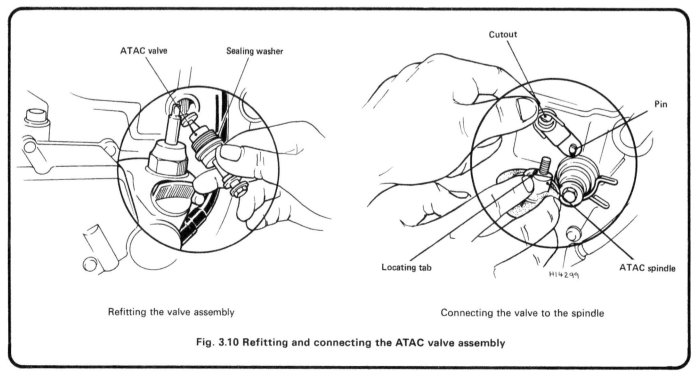

Refitting the valve assembly

Connecting the valve to the spindle

Fig. 3.10 Refitting and connecting the ATAC valve assembly

13 Engine lubrication system: general

Refer to Routine Maintenance for details of all regular adjustments and checks which must be carried out on the engine lubrication system; these include topping up the oil tank, cleaning the tank filter, checking the feed pipes and adjusting the pump control cable to synchronise the pump with the carburettor.

The low oil level warning system, comprising the tank-mounted switch, the warning lamp in the instrument panel and the related wiring, is electrically-operated and is described in Chapter 7.

14 Oil pump: removal, examination and refitting

1   The oil pump is mounted on the crankcase right-hand cover and is removed as described in Section 7 of Chapter 1, refitting being described in Section 35 of the same Chapter. The pump control cable and feed pipes are disconnected and connected as described in Sections 4 and 37.

2   When the oil pump has been removed, carefully examine it for damage to the main body casting and for correct operation of the control lever. Check that the O-ring around its fitting spigot is in good condition. Replace it if necessary.

3   If there are any signs of damage, or if you have good reason to suspect that the pump is not working properly, discard it and fit a new one. Stripping the pump for repair is not a practical proposition as there are no parts available to recondition it.

15 Engine lubrication system: the bleeding procedure

1   Bleeding must be carried out whenever any part of the system is disturbed, if the oil in the tank has been allowed to drain completely or run dangerously low, or if you have any reason to suspect the presence of air in the system. Air in the system will rapidly produce an airlock which will interrupt the constant supply of oil, resulting in severe engine damage due to the consequent loss of lubrication.

2   The bleeding operation consists of two parts, removing air from the oil tank, the oil tank/oil pump feed line, and from the pump itself, then removing air from the delivery side of the pump and from the oil pump/inlet stub feed line.

3   To complete the first part, thoroughly clean the oil pump and the area of crankcase around it and pack clean rag around the base of the pump. Check that the oil tank is full of oil, topping it up to the base of the filler neck if necessary. Slacken the hexagon-headed bleed screw which is situated on the top of the pump. Oil will flow from the orifice

thus exposed. Watch carefully until you can see no more air bubbles in the oil, check that no more bubbles are visible in the transparent feed line, and then replace and tighten down the bleed screw. Remove the rag and clean away the surplus oil.

4   This first part of the bleeding operation can be speeded up if the oil tank/oil pump feed line has been completely drained of oil or is full of air bubbles. Disconnect the feed line at the lower pump body union and allow the tank to drain into a clean container until the pipe is full of oil and all air bubbles have been expelled. This will fill the pipe much faster and flush out any air bubbles, if present, in the base of the tank, the oil filter assembly and the oil feed pipe itself. Once the feed line is full of oil, place a finger over the end temporarily to stop the flow of oil, reconnect it to the union on the pump body and secure the wire clip. Pour the drained oil, if totally clean, back into the tank and proceed to clear the pump of air as described in the previous paragraph. If this method is used, check that no air bubbles subsequently emerge in the feed line.

5   Once the oil tank, oil tank/oil pump feed line, and the oil pump itself have been cleared of air bubbles, proceed with the second part of the bleeding operation. This involves running the engine because the only way to expel air from the delivery side of the pump and from the oil pump/inlet stub feed line is to pump oil through, running the engine to operate the pump. The engine will therefore be running for a while with its normal supply of lubricant disconnected and an alternative method of lubrication must be used if the engine is not to suffer serious damage due to lack of lubrication.

6   Place the machine on its stand on level ground and in a well-ventilated area. This last is essential as excessive quantities of exhaust smoke and fumes will be produced. Check the oil level in the tank and top up to the base of the filler neck if necessary. Disconnect the petrol pipe at the tap and drain any petrol remaining in the petrol tank into a clean container. Mix up a small quantity (about a pint) of 25/50:1 petroil mixture and pour this into the petrol tank. Connect the petrol pipe up again and turn the petrol tap to the 'Res' position. Disconnect the oil pump/inlet stub feed line at the inlet stub union if this has not already been done. Start the engine and slowly warm it up until it will tick over smoothly. Using one finger, move the oil pump control lever around to the fully open position and keep it there. This will ensure that the pump is working at maximum capacity at idle speed and will hasten the process as much as is advisable. Keep the engine at as low a speed as possible, preferably at a constant idle. Watch the end of the oil feed pipe very carefully. If it has been completely emptied, it may take as long as 10 minutes for the oil to appear.

7   When oil does appear in the end of the pipe, continue the bleeding process until you are certain that no more air bubbles are appearing and that all traces of air have therefore been expelled. Due to the slow rate of delivery of the oil pump (less than 0.10 cc/min at idle speed) this operation is very time consuming and it might help to speed up the operation by wedging open the oil pump control lever with a piece of

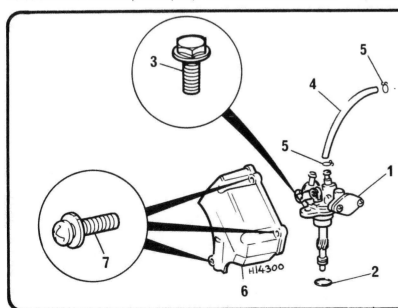

Fig. 3.11 Oil pump

1   Oil pump
2   O-ring
3   Bolt
4   Oil pipe
5   Clip – 2 off
6   Oil pump cover
7   Screw – 3 off

wood; if this is done carefully the control lever will remain in the fully open position, leaving your hands free for other tasks. The operation must, however, be carried out very carefully if the engine is not to suffer severe and premature damage due to the lack of lubrication which would result were air permitted to remain in the injection system.

8 As soon as the oil flowing from the end of the feed pipe is completely free of air bubbles, reconnect the oil feed pipe to the inlet stub union, secure its wire clamp, and stop the engine. Top up the oil tank to just below the filler neck and replace the filler cap. Ensure that all oil pipe unions are plugged firmly into place and that they are secured by their wire clips. Check that the oil lines are routed correctly and secured by such clamps or cable ties as are provided for this purpose. Wash off any surplus oil and check that the rag has been removed from the base of the oil pump. Drain any remaining petroil mixture from the petrol tank and replace it with fresh petrol. Connect the petrol pipe up and secure it with the wire clip. Check that the oil pump control lever has returned to the fully closed position and then fully open and close the throttle several times to settle the cable and to check that the pump control lever is operating correctly. Recheck the oil pump cable adjustment.

9 When taking the machine out on the road after bleeding the oil injection system, remember that the exhaust will be excessively smoky until the surplus oil has been used or burned up. Remember also to check for oil leaks in the system, these being readily apparent if the surplus oil was washed off as described. Any such leaks should be corrected immediately, before the machine is used further.

15.3 Removing oil pump bleed screw to purge system of air

Chapter 4 Ignition system

Contents

Specifications

Unless otherwise stated, information applies to all models

Ignition timing
Initial – @ idle speed to 3000rpm	19° BTDC – F mark aligned
Retard starts at	4000rpm
Full retard – @ 6000rpm	15.3 ± 2° BTDC

Alternator
Source coil resistance	157-213 ohm (Black/Red wire – earth)
Pulser coil resistance	100 ± 15 ohm (Blue/Yellow – Green/White wires)

Ignition HT coil
	MBX125	MTX125/200
Primary winding resistance	N/Av (Black/Yellow – Green)	0.2 – 0.4 ohm (Black/Yellow – centre pole)
Secondary winding resistance	3.6 – 4.6 ohm (HT lead – Green)	3.5 – 4.5 ohm (HT lead – centre pole)

Spark plug
	NGK	ND
Type – standard	BR9ES	W27ESR-U
Alternative hotter (softer) grade	BR8ES	W24ESR-U
Additional alternative – MTX125/200	BR7ES	W22ESR-U

Electrode gap:
MBX125, MTX125 RW-F, H, L, MTX200 RW-F	0.7 – 0.8 mm (0.028 – 0.032 in)
MTX125/200 RW-D	0.6 – 0.7 mm (0.024 – 0.028 in)

Torque wrench settings
Component	kgf m	lbf ft
Alternator rotor retaining nut	6.0 - 7.0	43 - 50.5
Pulser coil mounting bolts	0.8 - 1.2	6 - 9

1 General description

The ignition system is of the CDI (Capacitor Discharge Ignition) type, being powered by a source coil built into the generator stator. Power from this coil is fed directly to the CDI unit mounted beneath the frame, where it passes through a diode which converts it to direct current (dc). The charge is stored in a capacitor at this stage.

The spark is triggered by the pulser assembly which is mounted on the crankcase. As the magnetic rotor passes the pulser coil a small alternating current (ác) pulse is induced. This enters the CDI unit where it is rectified by a second diode. The heart of the CDI unit is a component known as a thyristor. It acts as an electric switch, which remains off until a small current is applied to its gate terminal. This causes the thyristor to become conductive, and it will remain in this state until any stored charge has discharged through the primary windings of the coil.

The sudden discharge of low-tension energy through the coil's primary windings in turn induces a high tension charge in the secondary coil. It is this which is applied to the centre electrode of the sparking plug, where it jumps the air gap to earth, igniting the fuel/air mixture.

As engine speed rises, it becomes necessary for the timing of the ignition spark to be altered in relation to the crankshaft to allow sufficient time for the combustion of the air/fuel mixture to take place at the optimum position. This function is controlled by the voltage build-up time in the pulser, which varies with the speed of the engine, acting on the thyristor gate circuit. Rather unusually, the timing is retarded rather than advanced; a condition which should be remembered when checking the timing.

This Chapter describes the procedures necessary to trace faults and to test the various components; refer to Routine Maintenance for information on the spark plug.

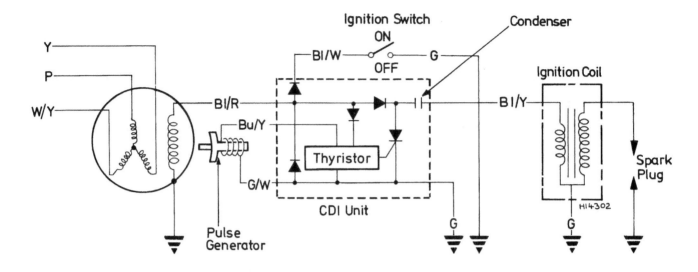

Fig. 4.1 Ignition circuit diagram

Colour key

Y	Yellow	BL/R	Black and red	G/W	Green and white	BL/Y	Black and yellow
P	Pink	BU/Y	Blue and yellow	BL/W	Black and white	G	Green
W/Y	White and yellow						

2 CDI system: fault diagnosis

1   As no means of adjustment is available, any failure of the system can be traced to the failure of a system component or a simple wiring fault. Of the two possibilities, the latter is by far the most likely. In the event of failure, check the system in a logical fashion, as described below.

2   Remove the spark plug, giving it a quick visual check noting any obvious signs of flooding or oiling. Fit the plug into the plug cap and rest it on the cylinder head so that the metal body of the plug is in good contact with the cylinder head metal. The electrode end of the plug should be positioned so that sparking can be checked as the engine is spun over using the kickstart.

3   *Important note.* The energy levels in electronic systems can be very high. **On no account** should the ignition be switched on whilst the plug or plug cap is being held. Shocks from the HT circuit can be most unpleasant. Secondly, it is vital that the plug is soundly earthed when the system is checked for sparking. The CDI unit **can be seriously damaged** if the HT circuit become isolated.

4   Having observed the above precautions, check that the kill switch is in the 'Run' position, turn the ignition switch to 'On' and kick the engine over. If the system is in good condition a regular, fat blue spark should be evident at the plug electrodes. If the spark appears thin or yellowish, or is non-existent, further investigation will be necessary. Before proceeding further, turn the ignition off and remove the key as a safety measure.

5   Ignition faults can be divided into two categories, namely those where the ignition system has failed completely, and those which are due to a partial failure. The likely faults are listed below, starting with the most probable source of failure. Work through the list systematically, referring to the subsequent sections for full details of the necessary checks and tests.

Total or partial ignition system failure:

    a)   Loose, corroded or damaged wiring connections, broken or shorted wiring between any of the component parts of the ignition system

    b)   Faulty main switch or engine kill switch

    c)   Faulty ignition HT coil

    d)   Faulty CDI unit

    e)   Faulty generator source coil

    f)   Faulty pulser coil

3 CDI system: checking the wiring

1   The wiring should be checked visually, noting any signs of corrosion around the various terminals and connectors. If the fault has developed in wet conditions it follows that water may have entered any of the connectors or switches, causing a short circuit. A temporary cure can be effected by spraying the relevant area with one of the proprietary de-watering aerosols such as WD40 or similar. A more permanent solution is to dismantle the switch or connector and coat the exposed parts with silicone grease to prevent the ingress of water. The exposed backs of connectors can be sealed off using a silicone rubber sealant.

2   Light corrosion can normally be cured by scraping or sanding the affected area, though in serious cases it may prove necessary to renew the switch or connector affected. Check the wiring for chafing or breakage, particularly where it passes close to part of the frame or its fittings. As a temporary measure damaged insulation can be repaired with PVC tape, but the wire concerned should be renewed at the earliest opportunity.

3   Using the wiring diagram at the end of the manual, check each wire for breakage or short circuits using a multimeter set on the resistance scale or a dry battery and bulb wired as shown in the accompanying illustration. In each case, there should be continuity between the ends of each wire.

4 Ignition and engine kill switches: testing

1   The ignition system is controlled by the ignition switch or main switch which is bolted to the fork top yoke. The switch has several terminals and leads, of which two are involved in controlling the

ignition system. These are the 'IG' terminal (black/white lead) and the 'E' terminal (green lead). The two terminals are connected when the switch is in the 'Off' position and prevent the ignition system from functioning by shorting the CDI unit to earth. When the switch is in the 'On' position the CDI/earth connection is broken and the system is allowed to function.

2   If the operation of the switch is suspect, reference should be made to the wiring diagram at the end of this book. The switch connections are also shown in diagrammatic form and indicate which terminals are connected in the various switch positions. The wiring from the switch can be traced back to the respective connectors where test connections can be made most conveniently.

3   The purpose of the test is to check whether the switch connections are being made and broken as indicated by the diagram. In the interests of safety the test must be made with the machine's battery disconnected, thus avoiding accidental damage to the CDI system or the owner. The test can be made with a multimeter set on the resistance scale, or with a simple dry battery and bulb arrangement as previously shown. Connect one probe lead to each terminal and note the reading or bulb indication in each switch position.

4   If the test indicates that the black/white lead is earthed irrespective of the switch position, trace and disconnect the ignition (black/white) and earth (green) leads from the ignition switch. Repeat the test with the switch isolated. If no change is apparent, the switch should be considered faulty and renewed. If the switch works normally when isolated, the fault must lie in the black/white lead between the switch and the CDI unit.

5   The kill switch, mounted on the right-hand handlebar, is tested in the same way as its connections and functions are exactly as described above.

6   If either switch is found to be faulty it must be renewed. While each is a sealed unit and can only, officially, be repaired by renewing it as a complete assembly; there is nothing to be lost by attempting to repair it if tests have proven it faulty. Depending on the owner's skill, worn contacts may be reclaimed by building up with solder or in some cases, merely cleaning with WD40 or a similar water dispersant spray.

5   Ignition HT coil: location and testing

1   The ignition HT coil is a sealed unit, and will normally give long service without need for attention. It is mounted beneath the frame top tube to the rear of the steering head, and is covered in use by the fuel tank. It follows that it will be necessary to remove the tank in order to gain access to the coil.

2   If a weak spark and difficult starting causes the performance of the coil to be suspect, it should be tested by a Honda service agent or an auto-electrical expert using the machine's CDI unit in conjunction with a special CDI tester to carry out a full check of the coil's performance. This is the only way the coil can be fully tested, but some idea of its condition can be gained from the tests below, and can be carried out by any owner who has a multimeter with the requisite measuring ranges.

3   First test the coil's primary winding resistance. Referring to the Specifications Section of this Chapter, make the meter connections between the wire terminals indicated (MBX125) or between the wire terminal and the coil centre pole/mounting lug (MTX125/200); while a specified resistance is not available for MBX125 models the value obtained should be similar to that given for the MTX125/200 models. Next measure the secondary winding resistance. On MBX125 models measure the resistance between the suppressor cap terminal and the green wire terminal; if the reading obtained is wrong, substitute a new suppressor cap and repeat the test to check that it is the coil at fault rather than the cap. On MTX125/200 models unscrew the HT lead retainer, withdraw the HT lead and measure between the HT terminal and the coil centre pole/mounting lug. Finally check that both windings are completely insulated from each other by measuring between the black/yellow wire terminal and the suppressor cap or HT terminal (as appropriate); if anything less than infinite resistance is measured, the coil is faulty.

4   Should any of these checks not produce the expected result, the coil should be taken to a Honda service agent or auto-electrician for a more thorough check. If the coil is found to be faulty, it must be renewed; it is not possible to effect a satisfactory repair.

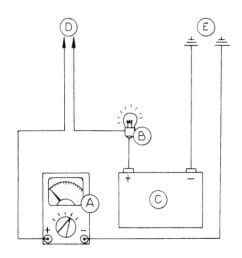

Fig. 4.2 Method of checking the wiring

A   Multimeter   D   Positive probe
B   Bulb   E   Negative probe
C   Battery

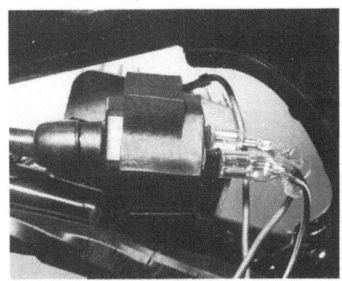

5.1a Location of ignition HT coil – MBX125 ...

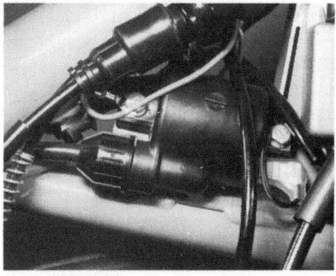

5.1b ... and MTX125/200

6 CDI unit: location and testing

1   The CDI unit takes the form of a sealed black box mounted on the frame top tubes, immediately in front of the fuel tank rear mounting, on MBX125 models and on the left-hand side of the frame, immediately behind the steering head, on MTX125/200 models. To remove the unit, withdraw the fuel tank as described in Chapter 3, unplug the wiring connector and disengage the unit from its rubber mounting.

2   To test the unit, Honda advise against the use of any test meter other than the Sanwa Electric Tester Type SP-10D (Honda part number 07308-0020000) or the Kowa Electric Tester (TH-5H), because the use of other instruments may result in inaccurate readings.

3   Most owners will find that they either do not possess a multimeter, in which case the CDI unit will have to be checked by a Honda Service Agent, or own a meter which is not of the specified make or model. In the latter case, a good indication of the unit's condition can be gleaned in spite of inaccuracies in the readings. If necessary, the CDI unit can be taken to a Honda Service Agent or auto-electrician specialist for confirmation of its condition.

4   The test details are given in the accompanying illustration in the form of a table of meter probe connections with the expected reading in each instance. If an ordinary multimeter is used, the resistance range may be determined by trial and error. The diagram illustrates the CDI unit connections referred to in the table. For owners not possessing a test meter the unit or the complete machine can be taken to a Honda Service Agent for testing.

5   If the CDI unit if found to be faulty as a result of these tests, it must be renewed. No repairs are possible. It might be worthwhile trying to obtain a good secondhand unit from a motorcycle breaker, in view of the cost of a new part.

6.1a CDI unit – MBX125 models

6.1b CDI unit – MTX125/200 models

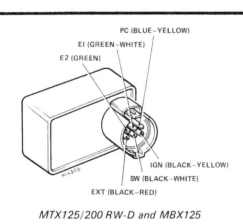

MTX125/200 RW-D and MBX125

MTX125 RW-F, H and MTX200 RW-F

⊕ ⊖	SW (Bl/W)	EXT (Bl/R)	PC (Bu/Y)	E1(G/W) E2(G)	IGN (Bl/Y)
SW(Bl/W)		∞	∞	∞	∞
EXT(Bl/R)	0.5-8		∞	∞	∞
PC(Bu/Y)	50-1000	50-1000		20-100	∞
E1(G/W) E2(G)	2-30	0.5-8	10-100		∞
IGN(Bl/Y)	∞	∞	∞	∞	

Fig. 4.3 Testing the CDI unit

Measuring ranges:
Sanwa tester (SP-10D) x k ohm
Kowa tester (TH-5H) x 100 ohm

7 Generator source coil: testing

The ignition source coil is tested exactly as described in Section 5 of Chapter 6, the reading to be expected from a sound coil being given in the Specifications Section of this Chapter. This also gives the meter connections necessary to make the test.

8 Pulser coil: location and testing

1 Although the pulser coil is a completely separate unit from the alternator stator, it cannot be removed independently unless a way can be found to slacken its wiring clamp retaining bolt without disturbing the alternator components. Refer to Sections 10 and 32 of Chapter 1 for instructions on removal and refitting.

2 To test the unit, remove the fuel tank as described in Chapter 2 and disconnect the coil wires (running up the frame front downtube) at the connector joining them to the main loom and measure the resistance between their terminals. If the unit is sound, a reading of 100 ± 15 ohms should be obtained. If the actual reading is significantly different, the unit is faulty and should be renewed.

3 Note however that if very little resistance is measured, indicating a short circuit between the two wires, or if very high resistance is encountered, indicating a broken wire, a fault may be due to the wires being trapped or damaged at some point along their length. Such faults may be traced and repaired easily by the private owner.

9 Spark plug (HT) lead and suppressor cap: examination

1 Erratic running faults and problems with the engine suddenly cutting out in wet weather can often be attributed to leakage from the high tension lead and spark plug cap. If this fault is present, it will often be possible to see tiny sparks around the lead and cap at night. One cause of this problem is the accumulation of mud and road grime around the lead, and the first thing to check is that the lead and cap are clean. It is possible to cure the problem by cleaning the components and sealing them with an aerosol ignition sealer, which will leave an insulating coating on both components.

2 Water dispersent sprays are also highly recommended where the system has become swamped with water. Both these products are easily obtainable at most garages and accessory shops. Occasionally, the suppressor cap or the lead itself may break down internally. If this is suspected, the components should be renewed.

3 Where the HT lead is permanently attached to the ignition coil it is recommended that the renewal of the HT lead is entrusted to an auto-electrician who will have the expertise to solder on a new lead without damaging the coil windings.

4 When renewing the suppressor cap, be careful to purchase one that is suitable for use with resistor spark plugs.

10 Checking the ignition timing

1 Since no provision exists for adjusting the ignition timing and since no ignition component is subject to mechanical wear, there is no need for regular checks; only if investigating a fault such as a loss of power or a misfire should the timing be checked.

2 The ignition timing can be checked only whilst the engine is running using a stroboscopic lamp; therefore a suitable timing lamp will be required. The inexpensive neon lamps should be adequate in theory, but in practice may produce a pulse of such low intensity that the timing mark remains indistinct. If possible, one of the more precise xenon tube lamps should be employed powered by an external source of the appropriate voltage. Do not use the machine's own battery as an incorrect reading may result from stray impulses within the machine's electrical system.

3 Remove the left-hand crankcase cover. Connect the timing lamp to the machine as directed by the lamp's manufacturer. Start the engine and aim the lamp at the generator rotor. Increase engine speed to 3000 rpm and check that the fixed index mark on the crankcase wall is aligned exactly with the 'F' mark on the rotor.

4 Check that the retard circuit is functioning correctly by increasing the engine speed slowly; at 4000 rpm the timing marks should begin to appear to move so that at 6000 rpm the full retard mark is now in alignment with the fixed index mark; note that while the full retard mark can be either two close-spaced parallel lines or a single line, it is easily identified as it is the only mark between the 'F' and the 'T' marks.

5 If there is any doubt about the ignition timing as a result of this check, take the machine to an authorised Honda dealer for an expert opinion. As already stated, there is no means of adjustment of the ignition timing on these machines, and a fault can be cured only by the testing of the source coil, pulser coil and CDI unit and by renewing any defective component. It is worth checking first that the pulser coil and stator plate are correctly positioned and securely fastened and that the rotor is located correctly on the crankshaft by its Woodruff key with its retaining nut securely fastened. Refer to Chapter 1.

10.4 Ignition timing retards as engine speed increases

Chapter 5 Frame and forks

Contents

Specifications

Unless otherwise stated, information applies to all models

Front forks

Travel:		
MBX125	150 mm (5.91 in)	
MTX125/200	230 mm (9.06 in)	
Stanchion maximum runout	0.02 mm (0.008 in)	
Spring free length:	**MBX125**	**MTX125 RW-F, H, L**
Standard	478.0 mm (18.8189 in)	604.0 mm (23.7794 in)
Service limit	468.0 mm (18.4252 in)	597.8 mm (23.5353 in)
Main spring free length:	**MTX125/200 RW-D**	**MTX200 RW-F**
Standard	504.0 mm (19.8425 in)	511.6 mm (20.1416 in)
Service limit	489.0 mm (19.2519 in)	506.5 mm (19.9409 in)
Top spring free length:		
Standard	101.6 mm (3.999 in)	85.7 mm (3.374 in)
Service limit	98.6 mm (3.8819 in)	84.0 mm (3.307 in)
Recommended fork oil	ATF (Automatic Transmission Fluid)	
Fork oil capacity:		
MBX125	139.0 ± 2.5 cc (4.89 ± 0.09 fl oz)	
MTX125/200 RW-D	311.5 ± 2.5 cc (10.97 ± 0.09 fl oz)	
MTX125 RW-F, H, L, MTX200 RW-F	352 cc (12.39 fl oz)	
Fork oil level – measured from top of stanchion, top plug and spring(s) removed, forks fully compressed:		
MBX125	200 mm (7.87 in)	
MTX125/200 RW-D	180 mm (7.09 in)	
MTX125 RW-F, H, L, MTX200 RW-F	N/Av	
Recommended air pressure	0 – 0.4 kg/cm² (0 – 5.7 psi)	
Maximum permissible pressure	3.0 kg/cm² (42.7 psi)	
Maximum difference between legs	0.1 kg/cm² (1.5 psi)	

Rear suspension

Wheel travel:	
MBX125	110 mm (4.33 in)
MTX125/200	200 mm (7.87 in)

Rear suspension unit

Spring free length:

MBX125	153.5 mm (6.0433 in)
Service limit	150.4 mm (5.9213 in)
MTX125/200 RW-D	190.0 mm (7.4803 in)
Service limit	186.2 mm (7.3307 in)
MTX125 RW-F, H, L	N/Av
Service limit	N/Av
MTX200 RW-F	N/Av
Service limit	N/Av

Torque wrench settings

Component	kgf m	lbf ft
Handlebar clamp bolts	1.8 – 3.0	13 – 22
Steering stem nut:		
MBX125	6.0 – 9.0	43 – 65
MTX125/200	8.0 – 12.0	58 – 87
Fork top plug	1.5 – 3.0	11 – 22
Top yoke pinch bolts	2.0 – 2.5	14.5 – 18
Bottom yoke pinch bolts:		
MBX125	2.4 – 3.0	17 – 22
MTX125/200	3.0 – 3.5	22 – 25
Damper rod Allen screw	1.5 – 2.5	11 – 18
Headlamp nacelle mounting bolts – MTX125/200	0.8 – 1.2	6 – 9
Pillion footrest locknuts – MTX125/200 RW-D	3.0 – 4.0	22 – 29
Swinging arm pivot bolt retaining nut:		
MBX125	5.5 – 7.0	40 – 50.5
MTX125/200	7.0 – 10.0	50.5 – 72
Rear suspension unit top mounting bolt:		
MBX125	3.0 – 4.0	22 – 29
MTX125/200	6.0 – 7.5	43 – 54
Rear suspension unit bottom mounting bolt:		
MBX125	3.0 – 4.0	22 – 29
MTX125/200	3.8 – 4.3	27.5 – 31
Rear suspension linkage pivot bolts or retaining nuts:		
Front arm/frame – MBX125	4.0 – 5.0	29 – 36
Front arm/frame – MTX125/200	6.0 – 7.5	43 – 54
Front arm/rear arm – MBX125	4.0 – 5.0	29 – 36
Front arm/rear arm – MTX125/200	6.0 – 7.5	43 – 54
Rear arm/swinging arm – MBX125	4.0 – 5.0	29 – 36
Rear arm/swinging arm – MTX125/200	9.0 – 12.0	65 – 87
Rear suspension unit bottom mounting – MTX125/200	6.0 – 7.5	43 – 54

1 General description

The front forks are of the coil spring, hydraulically-damped telescopic type, with each leg being fitted with an air valve in its top plug so that the fork's effective spring rate can be altered to suit the rider's needs. On MBX125 models, the stanchions bear on two Teflon-coated bushes which can be renewed individually to compensate for wear; on MTX125/200 models a single top bush only is fitted, which is not available as a separate part. Note that the forks fitted to the MTX125 RW-F, H, L and MTX200 RW-F models are different from those fitted to the earlier MTX125/200 RW-D models, incorporating larger diameter stanchions amongst other changes; components are not interchangeable between the two types.

The frame is a full-cradle structure fabricated from steel tubing and with welded joints.

The rear suspension is by Honda's 'Pro-Link' system in which the pivoted fork, or swinging arm acts on a single coil sprung, hydraulically-damped suspension unit via a two-piece linkage which alters progressively the rate at which the suspension unit spring is compressed. This provides compliant suspension at around the static load position and increasingly stiff suspension as the fork moves further throughout its travel.

2 Front fork legs: removal and refitting

1 Support the machine securely on level ground so that the front wheel is raised clear of the ground, then remove the front wheel as described in Chapter 6. Removing the retaining bolts where necessary, release the front brake hose or cable (as applicable) and the

speedometer cable from any clamps or guides securing them to the forks. On MBX125 models, remove its four mounting bolts and withdraw the mudguard; it may also be necessary to remove the fairing to provide working space.

2 Remove each valve cap and depress the valve core to release the air pressure. While the stanchions are still securely clamped in the yokes, slacken the threaded plug at the top of each leg; on MBX125 models a 17 mm open-ended spanner was found to fit closely the top of the plug squared section. Do not remove the top plug completely at this stage and take care not to damage the valve.

3 Slacken fully the pinch bolts securing the fork leg in both the top and bottom fork yokes and withdraw the leg by pulling it downwards out of the yokes. If corrosion obstructs removal of the leg, apply a liberal quantity of penetrating fluid, allow time for it to work, then release the leg by rotating it in the yoke as it is pulled downwards. In extremely stubborn cases, unscrew the air valve from each top plug, retighten the plugs (if slackened) and tap smartly on the top of each plug using a hammer and a wooden drift; take great care not to damage the plugs. Alternatively, if care is used it is permissible slightly to open up the split clamp of each yoke by removing the pinch bolt and working the flat blade of a screwdriver into the clamp. Exercise extreme caution when doing this, as the clamps can be overstressed and broken easily.

4 On refitting, check that the upper length of each stanchion is clean and polished, and that all traces of dirt and corrosion have been removed. Apply a thin smear of grease to ease the passage of the stanchion through the yokes and to prevent corrosion. When the stanchion is in position, ie with its top end flush with the upper surface of the top yoke so that the fork top plug is standing proud, tighten the pinch bolts just enough to retain the leg.

5 On MTX125/200 models, rotate both fork gaiters so that their 'Inside' markings face to the inside, ie next to the wheel, then tighten

their clamps securely so that their upper ends are against the bottom yoke. On MBX125 models refit the mudguard and tighten lightly the mounting bolts; check that in fully tightening these bolts the fork lower legs are not twisted out of line or stressed in any way.

6 Refit the front wheel but before final tightening of the various fasteners, push the machine off its stand and apply the front brake, then bounce the machine a few times to settle the front suspension components. Tighten first the front wheel spindle nut to the specified torque setting and then tighten the mudguard mounting bolts, where applicable, followed by the bottom and top yoke pinch bolts to the specified torque setting. It is essential to tighten the fasteners from the wheel spindle upwards so that the front forks are clamped securely but without stress.

7 Complete the refitting of the front wheel and (where applicable) the brake caliper components, then adjust the front brake and check that the front wheel is free to rotate, that the speedometer functions correctly, and that the front brake and front forks are working efficiently before taking the machine out on the road. Do not forget to set the fork leg air pressure to the required amount, as described in Routine Maintenance, and to refit the fairing (if removed) on MBX125 models.

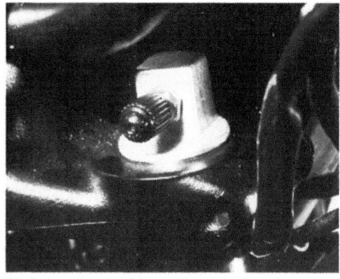

2.2a Remove valve cap and depress valve core to release air pressure before dismantling fork legs

2.2b Slacken top plug while fork leg is still clamped in the yokes, if fork legs are to be dismantled

2.3a Slacken pinch bolts in top ...

2.3b ... and bottom yokes to release fork legs

2.5 MBX125 – ensure that fork legs remain in alignment as mudguard mounting bolts are tightened

3 Front fork legs: dismantling and reassembly

1   Dismantle and rebuild the fork legs separately, and store the components of each leg in separate, clearly marked containers, so that there is no risk of exchanging components, thus promoting undue wear.

2   First slacken the damper rod Allen bolt at the bottom of each lower leg. If the damper rod is not locked, in this case by the pressure of the fork spring, it will rotate with the bolt. Tap smartly on the bolt head with a hammer and a suitable drift, then unscrew it. If it breaks free and can be unscrewed do not remove it completely at this stage; if it breaks free but merely rotates with the rod try compressing the fork leg (taking care to prevent damage to the air valve) against a solid object to apply more spring pressure. If all else fails proceed with the dismantling and use the more positive method of damper rod locking described below.

3   Remove the fork top plug. If this was not slackened while still in the machine, as described in the previous Section, clamp the stanchion upper end in a vice using padded jaw covers to prevent scratching. Unscrew the plug carefully; place a thickly-wadded piece of rag over the plug before it is finally removed. The plugs are under some pressure from the fork spring(s) and may be ejected with considerable force; use firm hand pressure to counter this until the plug threads are released.

4   Remove the fork spring(s) with the thick washer (where fitted), **noting carefully** which way up the spring(s) were fitted. Invert the leg over a container and tip out the fork oil; depress the leg slowly to expel as much oil as possible. On MTX125/200 models slacken its upper clamp and slide off the fork gaiter, then carefully prise the dust seal out of the lower leg and use a pair of circlip pliers to remove the large circlip beneath it. On MBX125 models carefully prise the dust seal out of the lower leg and use a pointed instrument to displace the wire circlip.

5   Remove the damper rod Allen bolt. If this has broken free from the lower leg but will not unscrew from the damper rod, obtain a piece of wooden dowel of the same length and diameter as the fork spring(s) and grind a coarse taper on one end. Compress the stanchion fully into

the lower leg and insert the dowel into the stanchion so that its tapered end engages with the head of the damper rod. Either rest the dowel outer end against a solid object so that the maximum pressure can be applied to lock the damper rod, clamp it in a vice, or clamp a self-locking wrench on to it so that an assistant can apply pressure and prevent rotation at the same time.

6   On MTX125/200 models, remove the stanchion from the lower leg and invert it to tip out the damper rod and rebound spring, then invert the lower leg to tip out the damper rod seat. The oil seal can be prised out of the lower leg using a large flat-bladed screwdriver which has had any sharp edges ground off; be very careful not to damage the top edge or to scratch the seal housing when using this method. **Do not apply excessive pressure**; if the seal proves difficult to remove, pour boiling water over the outside of the lower leg to expand the alloy sufficiently to release its grip on the seal. Withdraw the back-up ring beneath the seal.

7   On MBX125 models, clamp the fork lower leg by its spindle lug in a vice fitted with soft jaw covers, push the stanchion fully into the lower leg and pull it out as sharply as possible. Repeat the operation several times, using the slide hammer action of the bottom bush against the top bush to dislodge the oil seal. When the seal is released, the stanchion can be withdrawn complete with the seal and bushes. Thoroughly clean the stanchion upper end and smear it with fork oil to minimise damage as the seal, back-up ring and top bush are slid off. Invert the stanchion to tip out the damper rod and rebound spring, then invert the lower leg to tip out the damper rod seat. The stanchion bottom bush should not be disturbed unless absolutely necessary; to remove it carefully insert a thin screwdriver blade or similar into the vertical split in the bush, then spring the bush apart by just enough to ease it over the stanchion lower end.

8   On reassembly, all components should have been checked for wear and renewed as necessary and should be completely clean and dry.

9   On MBX125 models insert the damper rod, complete with piston ring and rebound spring, into the stanchion and press it down with the fork spring or wooden dowel so that it projects fully from the stanchion lower end, then refit the damper rod seat. Check that the bottom bush is correctly seated, smear the stanchion with fork oil and insert the

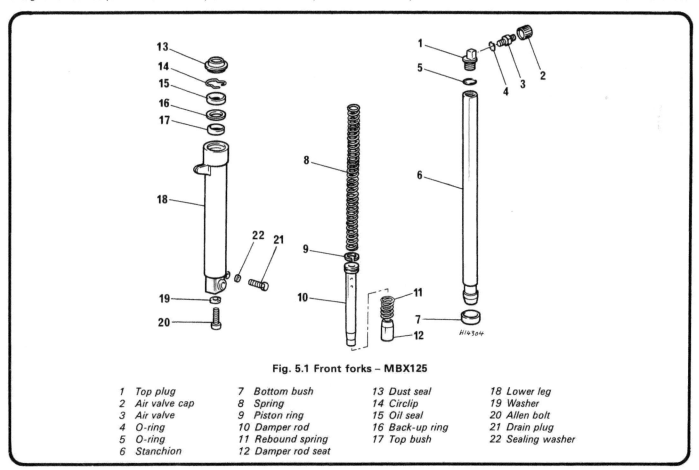

Fig. 5.1 Front forks – MBX125

1 Top plug	7 Bottom bush	13 Dust seal	18 Lower leg
2 Air valve cap	8 Spring	14 Circlip	19 Washer
3 Air valve	9 Piston ring	15 Oil seal	20 Allen bolt
4 O-ring	10 Damper rod	16 Back-up ring	21 Drain plug
5 O-ring	11 Rebound spring	17 Top bush	22 Sealing washer
6 Stanchion	12 Damper rod seat		

assembly into the fork lower leg. Press the stanchion fully into the lower leg to centralise the damper rod seat, apply a few drops of thread locking compound to the threads of the Allen bolt and refit it, not omitting its sealing washer. Lock the damper rod using the method employed on dismantling and tighten the bolt to a torque setting of 1.5 – 2.5 kgf m (11 – 18 lbf ft), then remove the spring or dowel. Smearing it with fork oil, slide the top bush down over the stanchion, followed by the back-up ring, and press the bush fully into its recess in the fork lower leg. Use a hammer and drift to tap the back-up ring and bush fully into place. Smear the stanchion with fork oil to protect the seal lip and slide down a new oil seal with its marked surface upwards. Press the seal squarely into the fork lower leg as far as possible by hand only, then use a length of tubing, as shown in the accompanying photograph, to push the seal into place until its retaining circlip groove is exposed. The tubing should be of the same outside diameter as the seal and should bear only on the seal's hard outer edge; check also that there are no burrs or sharp raised edges which might damage the seal. Firm hand pressure should suffice to fit the seal; if it is tapped into place with a hammer do not risk disturbing the seal by forcing it any further into the lower leg. As soon as the circlip groove is exposed, withdraw the tube and refit the seal retaining circlip. Pack grease above the seal for additional protection and press the dust seal firmly into place.

10 On MTX125/200 models clamp the lower leg vertically in a vice by its spindle lug and refit the back-up ring with its raised outer edge upwards. Coat the inner and outer diameters of the new seal with the recommended fork oil and push the seal square squarely into the bore of the fork lower leg by hand. Ensure that the seal is fitted squarely, then tap it fully into position, using a hammer and a suitably sized drift such as a socket spanner, which should bear only on the seal's hard outer edge. Tap the seal into the bore of the lower leg just enough to expose the circlip groove. Refit the retaining circlip securely in its groove. Refit the damper rod assembly to the stanchion, pushing it down with the spring(s) or wooden dowel, then place the damper rod seat over the damper rod end, using a smear of grease to stick it in place. Smear the sliding surface of the stanchion with a light coating of fork oil and carefully insert the stanchion into the lower leg, taking great care not to damage the sealing lips of the oil seal. Check that the threads of the damper rod bolt are clean and dry, apply a few drops of thread locking compound and fit the damper rod bolt. Do not forget the sealing washer fitted under the head of the bolt. Tighten the bolt only partially at first, using an Allen key of suitable size. Maintain pressure on the head of the damper rod and push the stanchion firmly as far down into the lower leg as possible to centralise the damper rod in the stanchion. The damper rod bolt can then be tightened to a torque setting of 1.5 – 2.5 kgf m (11 – 18 lbf ft) and the dowel or spring(s) removed. Pack grease above the seal for additional protection and press the dust seal firmly into place. Refit the gaiter, smearing grease over the stanchion upper end to prevent corrosion, but do not tighten the upper clamp yet.

11 On all models, clamp the lower leg by its spindle lug in a vice, using

padded jaws to prevent marking. Ensure that the leg is vertically upright and push the stanchion fully into the lower leg. Using a finely graduated vessel to ensure accuracy, pour in all but the last 10 – 20 cc of the specified amount of fork oil, then pump the stanchion slowly up and down several times to distribute the oil around the fork components and to expel as much air as possible. Press the stanchion fully into the lower leg, allow the oil to settle and measure the distance from the top of the stanchion to the top of the oil, using a dipstick as described in Routine Maintenance.

12 Add (or remove) oil as necessary until the oil level is exactly correct. **Note: it is essential** that the oil level is **exactly** the same in both fork legs and that it is at exactly the correct setting. As will be seen from the fact that only an approximate quantity is specified, the amount of fork oil is of secondary importance and should be varied as necessary to achieve the correct level.

13 With the oil level correct, refit the fork spring(s) with the washer between the two where applicable. Note that the springs should be refitted with their tapered-in coils downwards; where multi-rate springs are used, these are usually refitted with the closer-pitched coils upwards. Use the notes made on dismantling to confirm this. Refit the top plug, tightening it securely and if possible to the specified torque setting.

14 Depress the fork leg several times to ensure that it is working smoothly and correctly. The air pressure can be set, as described in Routine Maintenance, at this stage.

3.4a Note which way up the fork springs are fitted on removal

3.4b MBX125 – prise dust seal out of fork lower leg ...

3.4c ... then remove the oil seal retaining circlip

3.7 Removing the bottom bush – MBX125 – do not disturb unless necessary

3.9a Fork leg reassembly MBX125 – insert damper rod, with piston ring and rebound spring, into stanchion upper end ...

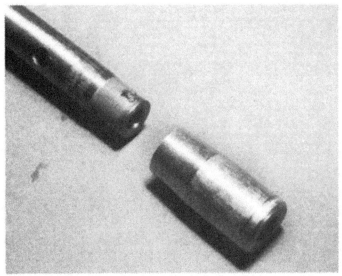

3.9b ... place damper rod seat over damper rod end ...

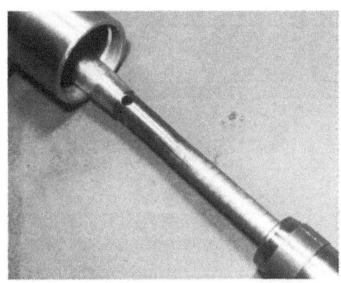

3.9c ... and insert stanchion assembly into lower leg – oil liberally

3.9d Tighten securely damper rod Allen bolt

3.9e Oil top bush before sliding into place

3.9f Refit seal back-up ring and tap carefully into lower leg – do not damage top bush

3.9g Smear fork oil over seal lips and stanchion to avoid damage on refitting

3.9h Press seal into place using suitable length of tubing ...

3.9i ... until groove is exposed so that circlip can be refitted

3.9j Pack grease above oil seal before refitting dust seal

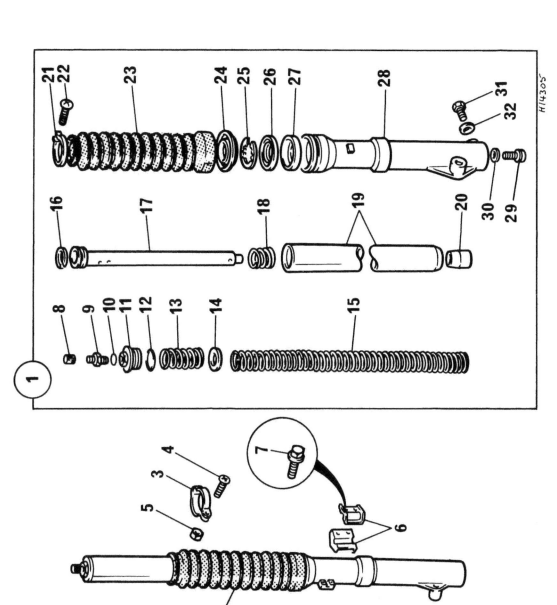

Fig. 5.2 Front forks – MTX125 and 200 RW-D

1 Right-hand fork leg
2 Left-hand fork leg
3 Speedometer cable clamp
4 Screw
5 Nut
6 Front brake cable clamp
7 Bolt – 2 off

8 Air valve cap – 2 off
9 Air valve – 2 off
10 O-ring – 2 off
11 Top plug – 2 off
12 O-ring – 2 off
13 Top spring – 2 off
14 Washer – 2 off

15 Main spring – 2 off
16 Piston ring – 2 off
17 Damper rod – 2 off
18 Rebound spring – 2 off
19 Stanchion – 2 off
20 Damper rod seat – 2 off
21 Clamp – 2 off

22 Screw – 2 off
23 Gaiter – 2 off
24 Dust seal – 2 off
25 Circlip – 2 off
26 Oil seal – 2 off
27 Back-up ring – 2 off

28 Right-hand lower leg
29 Allen bolt – 2 off
30 Washer – 2 off
31 Drain plug – 2 off
32 Sealing washer – 2 off

H1430S

134

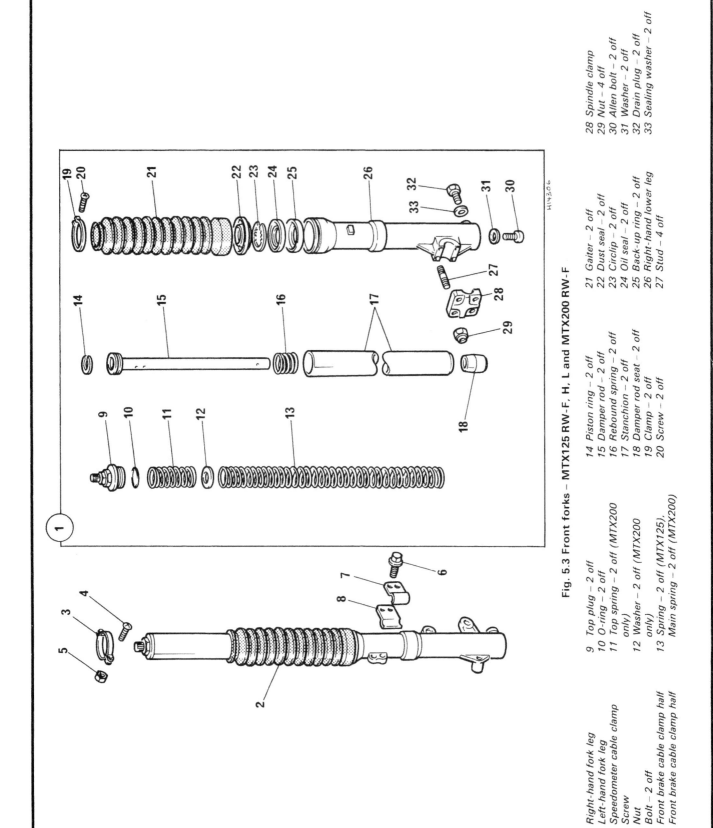

Fig. 5.3 Front forks – MTX125 RW-F, H, L and MTX200 RW-F

1 Right-hand fork leg
2 Left-hand fork leg
3 Speedometer cable clamp
4 Screw
5 Nut
6 Bolt – 2 off
7 Front brake cable clamp half
8 Front brake cable clamp half

9 Top plug – 2 off
10 O-ring – 2 off
11 Top spring – 2 off (MTX200 only)
12 Washer – 2 off (MTX200 only)
13 Spring – 2 off (MTX125), Main spring – 2 off (MTX200)

14 Piston ring – 2 off
15 Damper rod – 2 off
16 Rebound spring – 2 off
17 Stanchion – 2 off
18 Damper rod seat – 2 off
19 Clamp – 2 off
20 Screw – 2 off

21 Gaiter – 2 off
22 Dust seal – 2 off
23 Circlip – 2 off
24 Oil seal – 2 off
25 Back-up ring – 2 off
26 Right-hand lower leg
27 Stud – 4 off

28 Spindle clamp
29 Nut – 4 off
30 Allen bolt – 2 off
31 Washer – 2 off
32 Drain plug – 2 off
33 Sealing washer – 2 off

4 Front fork legs: examination and renovation

1   Carefully clean and dry all the components of the fork leg. Lay them out on a clean work surface and inspect each one, looking for excessive wear, cracks, or other damage. All traces of oil, dirt, and swarf should be removed, and any damaged or worn components renewed.

2   Examine the sliding surface of the stanchion or bushes, as applicable, and the internal surface of the lower leg, looking for signs of scuffing which will indicate that excessive wear has taken place. Slide the stanchion into the lower leg so that it seats fully. Any wear present will be easily found by attempting to move the stanchion backwards and forwards, and from side to side, in the bore of the lower leg. It is inevitable that a certain degree of slackness will be found, especially when the test is repeated at different points as the stanchion is gradually withdrawn from the lower leg, and it is largely a matter of experience to assess the amount of wear necessary to justify renewal of either the stanchion, the bushes, or the lower leg. It is recommended that the two components be taken to a motorcycle dealer for an expert opinion to be given if there is any doubt about the degree of wear found. Note that while wear zwill only become a serious problem after a high mileage has been covered, it is essential that such wear be rectified by the renewal of the components concerned if the handling and stability of the machine are impaired.

3   On MBX125 models a further check of bush wear is as follows. On examining their bearing surface (ie the top bush inside diameter and the bottom bush outside diameter), if the Teflon is worn away so that the copper material appears over more than $3/4$ of the whole bearing surface, that bush must be considered worn out. It would be best to renew all the bushes together to preserve equal fork performance.

4   Check the outer surface of the stanchion for scratches or roughness; it is only too easy to damage the oil seal during reassembly if these high spots are not eased down. The stanchions are unlikely to bend unless the machine is damaged in an accident. Any significant bend will be detected by eye, but if there is any doubt about straightness, roll down the stanchion tubes on a flat surface such as a sheet of plate glass. If the stanchions are bent to the point where the tubing is creased, they must be renewed, Unless specialised repair equipment is available it is rarely practicable to effect a satisfactory repair although slightly bent stanchions may be straightened.

5   Check the stanchion sliding surface for pits caused by corrosion. Such pits should be smoothed down with fine emery paper and filled if necessary, with Araldite. Once the Araldite has set fully hard, use a fine file or emery paper to rub it down so that the original contour of the stanchion is restored. Damage of this nature can be eliminated by the fitting of gaiters (MBX125 only); these are available from any good motorcycle dealer.

6   After an extended period of service, the fork springs may take a permanent set. If the spring lengths are suspect, then they should be measured and the readings obtained compared with the lengths given in the Specifications Section of this Chapter. It is always advisable to fit new fork springs where the length of the original items has decreased by a significant amount. Always renew the springs as a set, never separately.

7   If oil changes have not improved the situation and the damping is felt to be faulty, renew the damper rod piston ring and check that the rebound spring is in good condition, also that the damper rod passages are clean and unblocked.

8   Closely examine the dust seal and/or the gaiter (where fitted) for splits or signs of deterioration. If found to be defective, it must be renewed as any ingress of dirt will rapidly accelerate wear of the oil seal and fork stanchion. It is advisable to renew any gasket washers fitted beneath bolt heads as a matter of course. The same applies to the O-rings fitted to the fork top plugs and especially to the fork seals themselves.

9   The fork yokes are only likely to suffer damage of any sort in an accident; check carefully that there are no cracks, damaged or worn threads or any other signs of damage or wear. If they have been distorted by an accident this will usually be obvious to the eye and can only be cured by the renewal of the damaged item. Do not attempt to straighten the fork yokes as this may subsequently produce fatigue-induced cracks due to the stresses imposed. Check that the yokes are straight by pushing the fork stanchions temporarily into position; both should slide through the yokes with reasonable ease once the mating surfaces have been cleaned.

4.3 MBX125 – check bearing surface of both bushes and renew if excessively worn

4.7 Renew piston ring if damping is faulty – check rebound spring and damper rod passages

5 Steering head assembly: removal and refitting

1   Working as described in Section 2 of this Chapter, remove the front wheel, the mudguard (MBX125 only) and the front forks, then remove the seat, the side panels and the fuel tank.

2   On MBX125 models, remove the fairing, then remove its two mounting bolts and withdraw the headlamp assembly, unplugging the bulb connectors to release it. Disconnect all wires at the multi-pin block connector and unscrew the instrument drive cable retaining sleeve nuts to release the cables from the instruments. Remove its two retaining screws and withdraw the ignition switch cover from the fork top yoke, then remove the two bolts which secure the instrument mounting bracket to the top yoke. Check that all wires are disconnected which lead to the horn, the ignition switch and the handlebar switches, then withdraw the instrument assembly lifting the bracket bottom mountings out of their grommets in the bottom yoke.

3   On MTX125/200 models, remove the headlamp nacelle, then remove its two mounting bolts and withdraw the headlamp assembly, unplugging the bulb connectors to release it. Disconnect all wires at the multi-pin block connector and unscrew the instrument drive cable

retaining sleeve nuts to release the cables from the instruments, then check that all wires are disconnected which lead to the ignition switch, the horn, the handlebar switches and to the front turn signal lamps. Remove its four mounting bolts, noting the way in which the two small brackets are mounted, then withdraw the instrument mounting bracket complete with the instruments.

4 Remove the four handlebar clamp bolts and the two clamps, noting that on MBX125 models only, a small plastic plug is fitted in each bolt head and must be prised out before the bolt can be unscrewed. Move the handlebars backwards as far as possible clear of the steering head area without straining the control cables (and brake hose, where fitted) or wiring and secure them. Ensure that the front brake master cylinder reservoir is upright to avoid spillage. If the fuel tank was not removed it should be covered with a thick layer of old blanket or similar to avoid damaging the paintwork.

5 The yokes should now be cleared and ready for removal; the ignition switch can remain in place on the top yoke and the horn and front mudguard (MTX125/200 models only) can remain on the bottom yoke. These components need not be disturbed unless they or the yokes are to be renewed or repaired.

6 Remove the large nut and its washer from the centre of the top yoke. Using a soft-faced hammer, give the top yoke a gentle tap to free it from the steering head and lift it from position. Carry out a final check around the bottom yoke to ensure that all components have been removed which might prevent its release.

7 Support the weight of the bottom yoke and, using a C-spanner, of the correct size, remove the steering head bearing adjusting nut. If a C-spanner is not available, a soft metal drift may be used in conjunction with a hammer to slacken the nut.

8 Remove the top cone. The bottom yoke, complete with steering stem, can now be lowered from position. Ensure that any balls that fall from the bearings as the bearing races separate are caught and retained. It is quite likely that only the balls from the lower bearing will drop free, since those of the upper bearing will remain seated in the bearing cup. Carefully remove and count all the balls to ensure that none are lost (21 each, top and bottom, MBX125, 18 each top and bottom, MTX125/200), then place them in a container of solvent to clean them prior to checking for wear.

9 To remove the top and bottom bearing cups, if necessary, pass a long drift through the inner bore of the steering head and drive out the cup from the opposite end. The drift must be moved progressively around the race to ensure that it leaves the steering head evenly and squarely.

10 The lower of the two cones fits over the steering stem and may be removed by carefully drifting it up the length of the stem with a flat-ended chisel or a similar tool. Again, take care to ensure that the cone is kept square to the stem. There is a rubber seal and a metal washer fitted beneath the bottom cone; the seal must be renewed if damaged or worn to prevent the entry of dirt into the bearing.

11 Fitting of the new cups and cone is a straightforward procedure whilst taking note of the following points. Ensure that the cup locations within the steering head are clean and free of rust; the same applies to the steering stem. Lightly grease the stem and head locations to aid fitting and drift each cup or cone into position whilst keeping it square to its location. Fitting of the cups into the steering head will be made easier if the opposite end of the head to which the cup is being fitted has a wooden block placed against it to absorb some of the shock as the drift strikes the cup. Ensure that the drift used to fit the cups bears only on their outer edges (a socket spanner of the correct size would be best), while a tube of the same diameter as its raised inner edge should be used to refit the bottom cone; do not risk damaging the polished surface of the bearing track and do not forget to refit the metal washer and rubber seal to the steering stem before refitting the cone.

12 When positioning the balls, note that twenty-one are fitted in each race on MBX125 models, eighteen in each race on MTX125/200 models; this arrangement will leave a gap for one more ball which must not be fitted. If the balls do not have sufficient clearance they will skid against each other and wear at a much faster rate than normal.

13 Pack the bottom cone with grease and smear a liberal amount over the steering stem, the inside of the steering head lug and over both bearing cups. Use the grease to retain the balls as they are placed in the bottom cone and top cup, pack more grease over the balls and pass the steering stem up through the steering head. Be very careful not to dislodge any balls from either race.

14 Hold the bottom yoke firmly against the steering head, pack grease

into the top race and refit the top cone, taking care not to dislodge any balls, then refit the adjusting nut, tightening it firmly by hand pressure only. Rotate the bottom yoke throughout its full movement to ensure that it is free to move smoothly with no traces of stiffness, notchiness or free play. Slacken the nut until pressure is just removed.

15 To provide the initial setting for steering head bearing adjustment, tighten the adjusting ring carefully until resistance is felt then loosen it by $1/8$ to $1/4$ of a turn. Remember to check that the adjustment is correct, as described in Routine Maintenance, when the steering head assembly has been reassembled and the forks and front wheel refitted. Refit the fork yoke, followed by the washer and the steering stem nut; do not tighten the steering stem nut to its specified torque setting until the yokes have been fully aligned by refitting the fork legs and front wheel.

16 Finally, whilst refitting and reconnecting all disturbed components, take care to ensure that all control cables, drive cables, electrical leads, etc are correctly routed and that reference is made to the list of torque wrench settings given in the Specifications Section of this Chapter and of Chapter 6. Check that all controls and instruments function correctly before taking the machine on the public highway.

17 When refitting the handlebars and controls, use the marks provided by Honda as a guide to correct alignment. The handlebars should be placed on the bottom yoke clamps so that the punch marks next to each knurled section are aligned with the clamp top edge. Fit the upper clamp with its punch-marked ends to the front then refit and tighten to a torque setting of 1.8 – 3.0 kgf m (13 – 22 lbf ft) the two front clamp bolts; lastly fit and tighten to the same torque setting the two rear clamp bolts. On MBX125 models, press the small caps into the bolt heads.

18 The twistgrip, switch assemblies, and lever clamps (as applicable) are replaced by aligning the splits in their clamps (or between the two halves) with the appropriate handlebar punch mark or by ensuring that the locating pin (where fitted) engages with the corresponding hole in the handlebar. Where fitted, refit the hydraulic master cylinder clamp with its arrow pointing upwards, then tighten first the top clamp bolt, followed by the bottom clamp bolt.

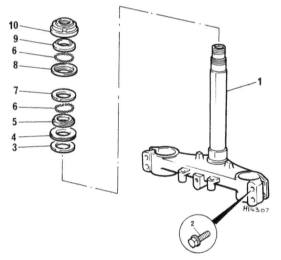

Fig. 5.4 Steering head assembly – typical

1 Bottom yoke
2 Pinch bolt – 4 off
3 Washer
4 Dust seal
5 Bottom cone
6 Steel ball
7 Bottom cup
8 Top cup
9 Top cone
10 Adjusting nut

6 Steering head bearings: examination and renovation

1 Before commencing reassembly of the steering head component parts, take care to examine each of the steering head bearings. The ball bearing tracks of their respective cup and cone bearings should be polished and free from any indentations or cracks. If wear or damage is evident, then the cups and cones must be renewed as complete sets.

2 Carefully clean and examine the balls contained in each bearing assembly. These should also be polished and show no signs of surface cracks or blemishes. If any one ball is found to be defective, then the

complete set should be renewed. Remember that a complete set of these balls is relatively cheap and it is not worth the risk of refitting items that are in doubtful condition. On MBX125 models, a total of forty-two $^3/_{16}$ in (No. 6) balls are used, while on MTX125/200 models a total of thirty-six $^1/_4$ in (No. 8) balls are fitted; in both cases the balls are divided equally between the top and bottom races.

3 Check that the seal at the bottom of the steering stem is in good condition. It keeps dirt and water out of the bearings and must be renewed if it is at all damaged or worn.

7 Frame: examination and renovation

1 The frame is unlikely to require attention unless accident damage has occurred. In some cases, renewal of the frame is the only satisfactory remedy if it is badly out of alignment. Only a few frame specialists have the jigs and mandrels necessary for resetting the frame to the required standard of accuracy, and even then there is no easy means of assessing to what extent it may have been over-stressed.

2 After the machine has covered a considerable mileage, it is advisable to examine the frame closely for signs of cracking or splitting at the welded joints. Rust corrosion can also cause weakness at these joints. Minor damage can be repaired by welding or brazing, depending on the extent and nature of the damage.

3 Remember that a frame which is out of alignment will cause handling problems and may even promote 'speed wobbles'. If misalignment is suspected, as a result of an accident, it will be necessary to strip the machine completely so that the frame can be checked, and if necessary, renewed.

8 Rear suspension: removal and refitting

1 **Note:** while it is possible, with a little care, to remove the pivot bolts and separate the suspension unit, linkage and swinging arm so that all three can be removed individually, it is recommended that the complete suspension be removed as a single assembly, as described below. This will permit all components to be checked for wear or damage and all bearings to be lubricated at the same time.

2 First remove the rear wheel as described in Chapter 6, then remove the seat, both side panels and the chainguard. If it is necessary to gain access to the suspension unit top mounting bolt, remove also the fuel tank. On MTX125/200 models unhook the brake pedal return spring from the swinging arm and remove the brake pedal pinch bolt so that the pedal shaft can be pulled out. Unhook the stoplamp rear switch spring and withdraw the brake pedal. On MTX200 RW-F models release its securing tie and remove the remote reservoir from its mountings; arrange the reservoir so that it will not foul any other component as the suspension is removed.

3 Remove the retaining nuts from the suspension unit top mounting, the linkage front arm/frame mounting and the swinging arm pivot bolt, then remove the bolts in the same order. If any bolt is difficult to remove because of dirt or corrosion, apply a liberal quantity of penetrating fluid and allow time for it to work before tapping out the bolt with a hammer and drift. Manoeuvring it carefully to avoid damage, lift out the swinging arm and rear suspension, disengaging it from the chain.

4 On reassembly, grease all bearings thoroughly and check that all components are in place, then manoeuvre the assembly into position, remembering to pass the swinging arm left-hand fork end through the chain. Refit the swinging arm pivot bolt, greasing it thoroughly and tighten its retaining nut to the torque setting specified. Check that it moves easily, with no traces of stiffness or free play, then align the linkage front arm with its frame mounting, grease the pivot bolt and refit it, followed by refitting the suspension unit top mounting bolt, suitably greased. Tighten their retaining nuts to the specified torque settings, but note that punch marks on the suspension unit bottom mounting and on the linkage rear arm (on the right-hand side) must be aligned before tightening the unit bottom mounting bolt. This is to ensure that the bonded rubber bush fitted at this point is tightened in approximately its normal working position to minimise distortion and to prevent premature wear.

5 Refit all components that were removed on dismantling and check for correct chain adjustment, suspension movement and rear brake operation before taking the machine out on the road.

8.2 Remove chainguard if necessary to provide working space

8.3a Remove suspension unit top mounting bolt ...

8.3b ... followed by the linkage front arm/frame pivot bolt (arrowed)

8.3c Remove swinging arm pivot bolt ...

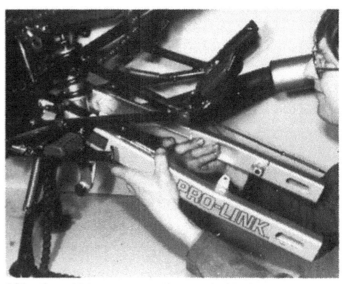

8.3d ... then withdraw rear suspension as a single unit

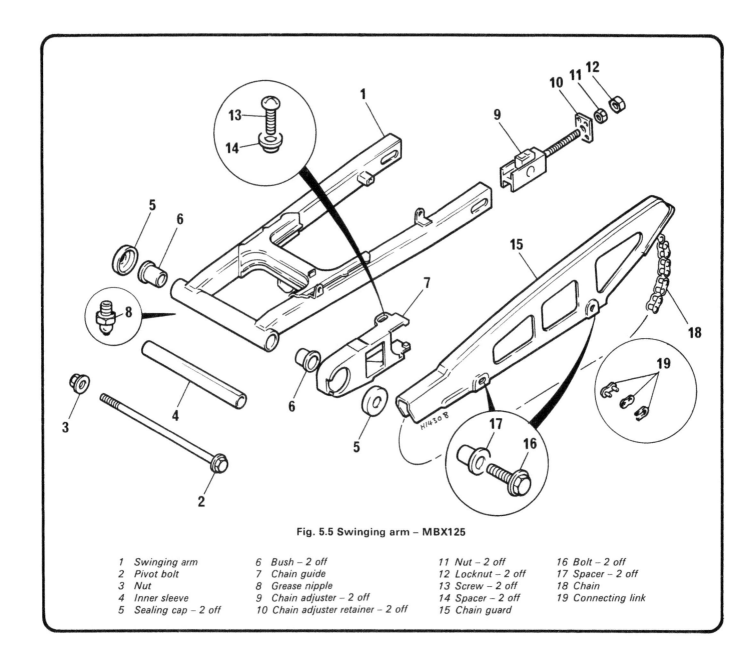

Fig. 5.5 Swinging arm – MBX125

1	Swinging arm	6	Bush – 2 off	11	Nut – 2 off	16	Bolt – 2 off
2	Pivot bolt	7	Chain guide	12	Locknut – 2 off	17	Spacer – 2 off
3	Nut	8	Grease nipple	13	Screw – 2 off	18	Chain
4	Inner sleeve	9	Chain adjuster – 2 off	14	Spacer – 2 off	19	Connecting link
5	Sealing cap – 2 off	10	Chain adjuster retainer – 2 off	15	Chain guard		

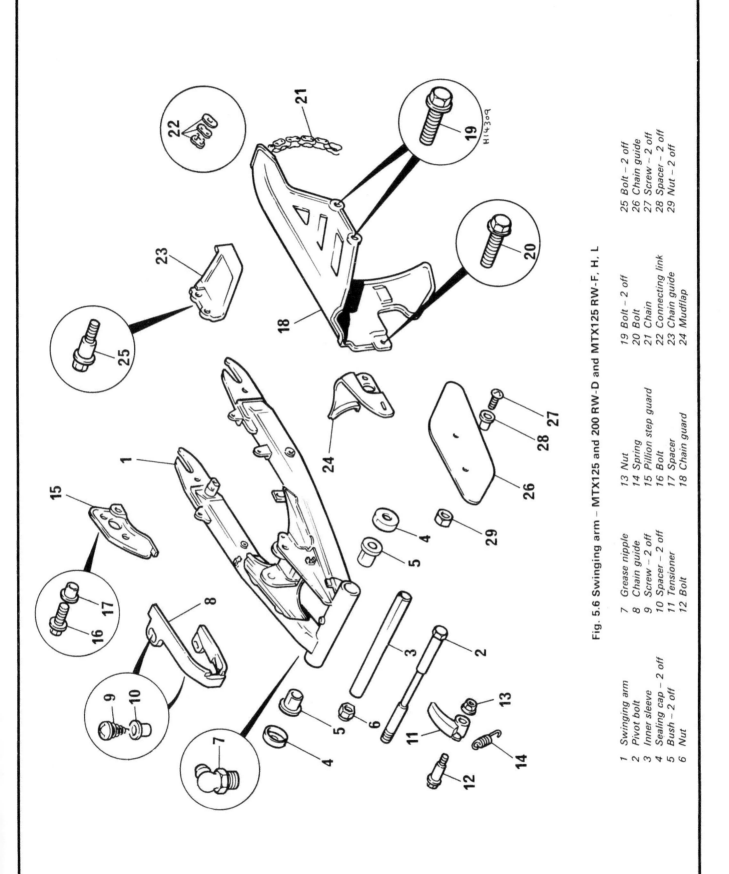

Fig. 5.6 Swinging arm – MTX125 and 200 RW-D and MTX125 RW-F, H, L

1 Swinging arm
2 Pivot bolt
3 Inner sleeve
4 Sealing cap – 2 off
5 Bush – 2 off
6 Nut

7 Grease nipple
8 Chain guide
9 Screw – 2 off
10 Spacer – 2 off
11 Tensioner
12 Bolt

13 Nut
14 Spring
15 Pillion step guard
16 Bolt
17 Spacer
18 Chain guard

19 Bolt – 2 off
20 Bolt
21 Chain
22 Connecting link
23 Chain guide
24 Mudflap

25 Bolt – 2 off
26 Chain guide
27 Screw – 2 off
28 Spacer – 2 off
29 Nut – 2 off

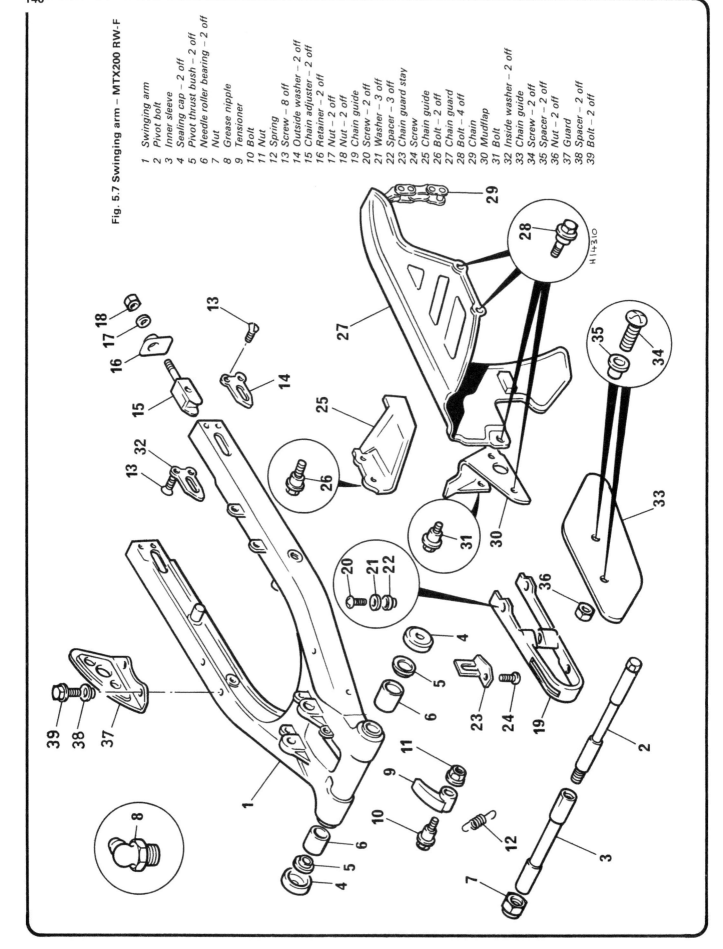

Fig. 5.7 Swinging arm – MTX200 RW-F

1 Swinging arm
2 Pivot bolt
3 Inner sleeve
4 Sealing cap – 2 off
5 Pivot thrust bush – 2 off
6 Needle roller bearing – 2 off
7 Nut
8 Grease nipple
9 Tensioner
10 Bolt
11 Nut
12 Spring
13 Screw – 8 off
14 Outside washer – 2 off
15 Chain adjuster – 2 off
16 Retainer – 2 off
17 Nut – 2 off
18 Nut – 2 off
19 Chain guide
20 Screw – 2 off
21 Washer – 3 off
22 Spacer – 3 off
23 Chain guard stay
24 Screw
25 Chain guide
26 Bolt – 2 off
27 Chain guard
28 Bolt – 4 off
29 Chain
30 Mudflap
31 Bolt
32 Inside washer – 2 off
33 Chain guide
34 Screw – 2 off
35 Spacer – 2 off
36 Nut – 2 off
37 Guard
38 Spacer – 2 off
39 Bolt – 2 off

9 Swinging arm and rear suspension linkage: examination and renovation

1 Before commencing work, check with a local Honda Service Agent exactly what is available for the machine being worked on. In some cases bushes are available only as part of a larger component.
2 Referring to the photographs and illustration accompanying the text, dismantle the assembly into its component parts and wash carefully each one in solvent to remove all traces of dirt and old grease. Components such as the chain guides, pillion footrests and linkage mud shields need not be removed unless necessary.
3 All pivot bearings are formed by a bush of metal or of synthetic material, or in some cases bonded rubber bushes or needle roller bearings, that are pressed into the bearing housing. A steel inner sleeve is fitted in the centre of this bearing and the pivot bolt passes through the sleeve. Sealing caps are fitted at each end of each bearing.
4 Wear will only be found, therefore, between the pivot bolt and inner sleeve and between the inner sleeve and bearing. After thorough cleaning, reassemble each bearing and feel for free play between the components. If excessive free play is felt, examine each component renewing any that shows signs of deep scoring or scuffing. Also check all components for wear due to the presence of dirt or corrosion, and check the swinging arm and linkage arms for signs of cracks, distortion or other damage, renewing any worn items.
5 Check that all pivot bolts are straight and unworn and that their threads are undamaged. Use emery paper to polish off all traces of corrosion and smear grease over them on reassembly.
6 The bushes or needle roller bearings can be removed either by tapping them out or by using a variation of a drawbolt arrangement as shown in the accompanying illustration. Note that since this will almost certainly damage the bush or bearing, they should not be disturbed unless renewal is necessary. On refitting, the drawbolt arrangement shown must be used to avoid the risk of damage. Clean the housing thoroughly and smear grease over the bearing and its housing to aid refitting. Needle roller bearings are refitted with their marked surfaces facing outwards.
7 The sealing caps should be examined closely and renewed if worn; they must be in good condition to exclude dirt and water from the bearings.
8 On reassembly, pack molybdenum disulphide grease (containing at least 45% molybdenum disulphide) into each bearing and sealing cap, and smear it over all inner sleeves and pivot bolts. Check that all grease nippes are clear by pumping grease into the bearings.
9 Refer to the accompanying photograph and illustrations on reassembly to ensure that all components are correctly refitted. Note

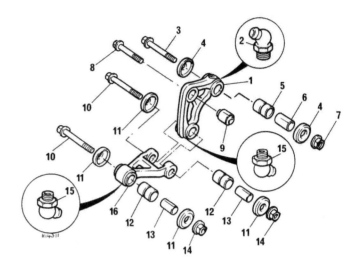

Fig. 5.8 Rear suspension linkage – typical

1 Linkage rear arm
2 Grease nipple
3 Bolt
4 Sealing cap – 2 off
5 Bush
6 Inner sleeve
7 Nut
8 Bolt
9 Suspension unit bottom mounting bush
10 Bolt – 2 off
11 Sealing cap – 4 off
12 Bush – 2 off
13 Inner sleeve – 2 off
14 Nut – 2 off
15 Grease nipple – 2 off
16 Linkage front arm

that the linkage front arm has the word 'Up' cast on one surface of its single larger mounting boss to show which way round it is fitted. Do not forget to refit the sealing caps to each bearing, ensuring that their lips seat correctly. All linkage pivot bolts are fitted from left to right on MBX125 models; on MTX125/200 models the swinging arm pivot bolt should be refitted from left to right, but all linkage pivot bolts are fitted from right to left, particularly the suspension unit bottom mounting bolt. Note that the unit bottom mounting threaded side must be on the left. Tighten all retaining nuts to the torque settings given in the Specifications Section of this Chapter.

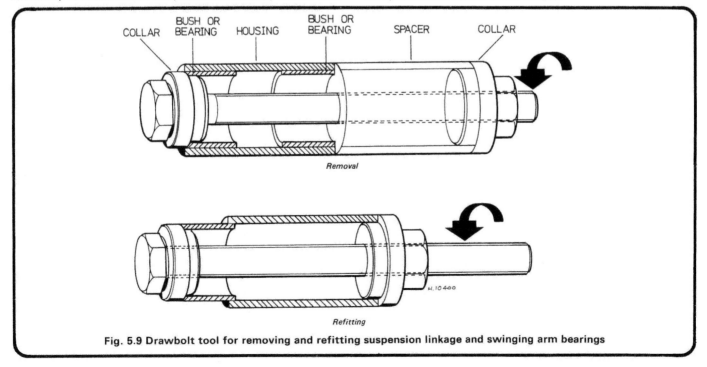

Fig. 5.9 Drawbolt tool for removing and refitting suspension linkage and swinging arm bearings

9.2 Remove pivot bolts to dismantle suspension – clean carefully each component

9.3a Pivot bearings are made up of a steel inner sleeve with an outer bush of synthetic material ...

9.3b ... or of metal – end caps incorporate seals to exclude dirt and water

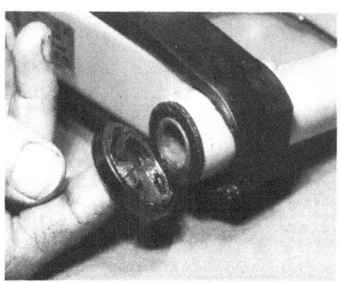

9.8 Pack bearings and caps with specified grease on reassembly

9.9a Ensure all components are correctly aligned ...

9.9b ... and that pivot bolts are correctly fitted and tightened

10 Rear suspension unit: removal and refitting

1   As mentioned in Section 8, the suspension unit is best removed as part of the rear suspension, but if it is necessary to detach the unit alone, proceed as follows.

2   Place the machine on a stand so that it is supported securely with the rear wheel clear of the ground, then remove the seat, the side panels and if necessary to reach the suspension unit top mounting bolt, the fuel tank. On MTX125/200 models remove the air filter casing mounting bolts to gain working space; it may be necessary to remove the filter casing but if a little ingenuity is exercised it should be possible to manoeuvre the unit from position without disturbing the filter casing. On MTX200 RW-F models, release the reservoir from its mountings. On all models, take the weight off the rear suspension by wedging a wooden block under the rear wheel, then remove its retaining nut and tap out the suspension unit top mounting bolt.

3   Exposing the unit bottom mounting may prove to be awkward; remove if necessary the mud shields and chain guides to gain extra working space, then either lift the wheel and support it on blocks as soon as the mounting bolt is accessible, or remove the linkage pivot bolts and lower the unit to the ground; it may be necessary to remove the linkage front arm to permit this. Be careful not to damage the unit.

4   Reassembly is a reversal of the dismantling procedure chosen. If the mounting bolts are to be renewed, use only genuine Honda parts to be sure of mountings of sufficient strength. Smear grease over the bolts and refit them from left to right on MBX125 models. On MTX 125/200 models, fit the bolts from right to left; note that the unit's bottom mounting threaded side must, therefore, be on the left. Tighten all bolts to the torque settings specified.

11 Rear suspension unit: dismantling, examination and reassembly

1   Note that the suspension unit is sealed for life and can only be renewed to rectify any faults such as if it is dented or damaged, if the damper rod is bent, or if any signs of oil leakage can be seen. To remove the spring, proceed as follows.

2   On MBX125 models some means must be found of compressing the spring safely so that the spring retainer can be pulled out. While this is usually achieved by placing the unit bottom mounting in a vice and having an assistant pull down on the spring, the safest method is to apply a pair of spring compressors to the spring coils. Compressors are readily available in auto accessory shops for use on the MacPherson strut front suspension of many cars, particularly Fords.

3   On MTX125/200 RW-D and MTX125 RW-F, H, L models, clamp the unit top mounting eye in a vice and compress the spring as described above, then use a pointed instrument to displace the spring top seat retaining circlip from its groove in the unit body; note carefully which groove is used of the three provided. With the spring still compressed, hold the unit bottom mounting and slacken its locknut, then unscrew the bottom mounting and the locknut. Withdraw the metal plate with the drain tube attached, followed by the spring bottom seat, the spring guide, the dust seal and the spring itself with its top collar and seat.

4   When it has been removed, measure the spring free length and renew it if it is settled to less than the service limit specified. Renew the damper assembly if any of the faults listed above are found, or if the damping action is weak.

5   The unit top mounting is a bonded rubber bush on MBX125 models and a metal bush with a steel inner sleeve on MTX125/200 models, check it for wear or damage and renew it if faulty. The bush is removed and refitted using a modified version of the drawbolt assembly described in Section 9 of this Chapter.

6   On MBX125 models reassembly is a straightforward reversal of the dismantling sequence. On MTX125/200 models the same applies but the following points should be noted. When refitting the top seat retaining circlip, note that while it should be refitted in the groove from which it was removed, a measure of preload adjustment can be achieved, if desired, by fitting the circlip in a different groove; the groove nearest the unit bottom mounting will give the hardest spring setting, this gradually softening as the circlip is moved up towards the unit top mounting.

7   Also on MTX125/200 models, when refitting the spring it must be compressed safely while the bottom mounting components are refitted in their correct positions. Apply thread locking compound to the damper rod threads, screw the locknut fully on to the damper rod, followed by the unit bottom mounting which should be tightened securely. Note that a torque setting of 6.0 – 7.5 kgf m (43 – 54 lbf ft) is specified and should be used if a means can be found of applying it. Slowly release spring pressure and rotate the metal plate until the pin projecting from it aligns with the cutout in the bottom mounting so that the drain tube will be in the correct position when the unit is refitted to the machine. Pack the unit top mounting with molybdenum disulphide grease before refitting the inner sleeve and check that the seals and sealing caps are correctly refitted.

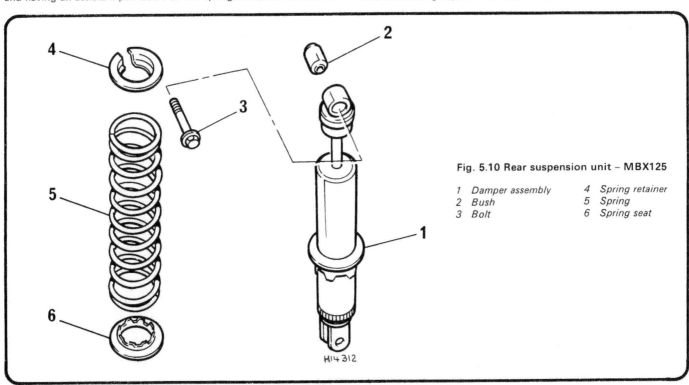

Fig. 5.10 Rear suspension unit – MBX125

1 Damper assembly	4 Spring retainer
2 Bush	5 Spring
3 Bolt	6 Spring seat

H14 312

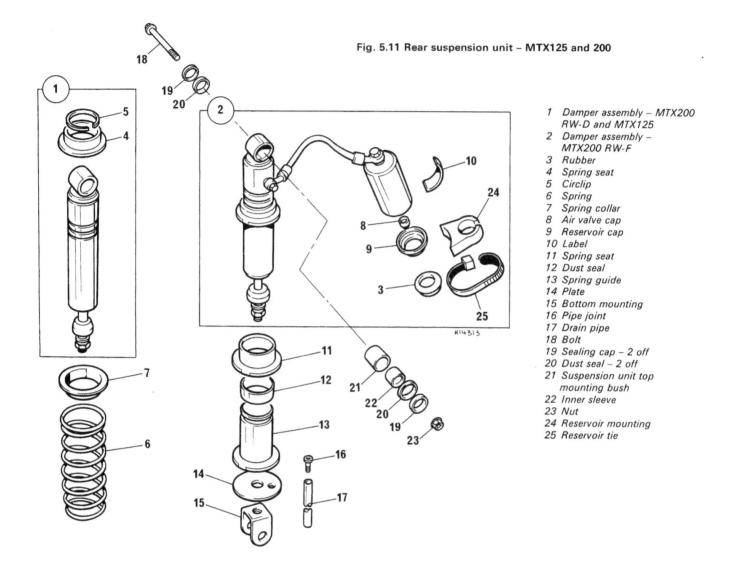

Fig. 5.11 Rear suspension unit – MTX125 and 200

1 Damper assembly – MTX200
  RW-D and MTX125
2 Damper assembly –
  MTX200 RW-F
3 Rubber
4 Spring seat
5 Circlip
6 Spring
7 Spring collar
8 Air valve cap
9 Reservoir cap
10 Label
11 Spring seat
12 Dust seal
13 Spring guide
14 Plate
15 Bottom mounting
16 Pipe joint
17 Drain pipe
18 Bolt
19 Sealing cap – 2 off
20 Dust seal – 2 off
21 Suspension unit top
   mounting bush
22 Inner sleeve
23 Nut
24 Reservoir mounting
25 Reservoir tie

12 Footrests, stands and controls: examination and renovation

1   At regular intervals all footrests, the stand, the brake pedal and the gearchange lever should be checked and lubricated. Check that all mounting nuts and bolts are securely fastened, using the recommended torque wrench settings where these are given. Check that any securing split pins are correctly fitted. On later MTX125/200 models check the condition of the side stand rubber pad and renew it if necessary, as described in Routine Maintenance.
2   Check that the bearing surfaces at all pivot points are well greased and unworn, renewing any component that is excessively worn. If lubrication is required, dismantle the assembly to ensure that grease can be packed fully into the bearing surface. Return springs, where fitted, must be in good condition with no traces of fatigue and must be securely mounted.
3   If accident damage is to be repaired, check that the damaged component is not cracked or broken. Such damage may be repaired by welding if the pieces are taken to an expert, but since this will destroy the finish, renewal is usually the most satisfactory course of action. If a component is merely bent it can be straightened after the affected area has been heated to a dull cherry red, using a blowlamp or welding

torch. Again the finish will be destroyed, but painted surfaces can be repainted easily, while chromed or plated surfaces can only be replated, if the cost is justified.

13 Speedometer and tachometer heads: removal, examination and refitting

1   On MBX125 models, remove the fairing, then remove its two mounting bolts and withdraw the headlamp assembly, unplugging the bulb connectors to release it. Unscrew the knurled sleeve nuts to release the drive cables from the instruments then remove the two chromed dome nuts and their washers to release the instrument panel from the mounting bracket; all bulbs may now be unplugged from the base of the instrument panel.
2   To remove the panel from the machine, unplug all the bulbs, noting where each is fitted, then remove the three retaining nuts and disconnect the temperature gauge wires from the unit; note which wire goes to which terminal before they are removed. Alternatively, all wires may be disconnected at the multi-pin block connector so that the panel can be withdrawn as a single unit. When all wires are disconnected, withdraw the instrument panel.
3   Remove the three screws and withdraw the panel top cover, then

remove the four screws from underneath the panel (one set in each corner) and separate the panel top half from the bottom. The speedometer and tachometer are each retained by two screws, while the temperature gauge may be withdrawn as soon as its terminals are disconnected. Reverse the dismantling procedure to reassemble and refit the instrument panel.

4   On MTX125/200 models, remove the headlamp nacelle then remove its two mounting bolts and withdraw the headlamp assembly, unplugging the bulb connectors to release it. Unscrew the knurled sleeve nuts to release the drive cables from the instruments, then remove the two dome nuts and their washers which secure each instrument to the mounting bracket; all bulbs may now be unplugged from the bases of the two instruments.

5   The tachometer can now be removed, but the temperature gauge wires must be disconnected by removing the retaining nut and washer at each terminal before the speedometer can be withdrawn; note which wire goes to which terminal before they are removed. Take care not to lose the spacers from the two mounting rubbers as each instrument is withdrawn.

6   The speedometer and tachometer on MTX125/200 models are sealed units which cannot be dismantled any further and are refitted by reversing the dismantling procedure.

7   Apart from defects in either the drive or drive cable, a speedometer or tachometer which malfunctions is difficult to repair. Fit a replacement or alternatively entrust the repair to a competent instrument repair specialist. A further alternative is to obtain a speedometer from a breaker who may have perfectly good instruments with scratched cases and glass etc.

8   Remember that a speedometer in correct working order is a statutory requirement in the UK. Apart from this legal necessity, reference to the odometer reading is the most satisfactory means of keeping pace with the maintenance schedules.

9   Note that these instruments are delicate and should be handled carefully at all times; do not allow them to drop. Do not hold them upside down or damping fluid may leak out and ruin them, and never allow dirt, oil, grease or water to get into them.

13.1a Unscrew chromed dome nuts to release instruments from mounting bracket ...

13.1b ... note different arrangement on MBX125

13.1c Instrument and warning bulbs may then be unplugged

13.3 MBX125 – instruments are each retained by two screws

14 Speedometer and tachometer drives: examination and renovation

Cables – all models

1   If an instrument fails to operate suddenly, or if its movement is jerky or sluggish, check that the cable is not broken, and remove the inner cable to check that it is adequately lubricated and not worn due to a trapped or kinked outer cable. The inner cables can be renewed individually, if necessary.

2   The cables should be removed at regular intervals so that they can be checked for wear or damage and so that the inner cables can be greased. Refer to Routine Maintenance.

3   The cables are secured at their upper ends by knurled sleeve nuts which should be slackened and tightened with a pair of pliers on removal and refitting. Their lower ends are pressed into housings in the front brake backplate, speedometer gearbox or crankcase right-hand cover (as applicable) and are retained by a single screw which engages with a groove in the outer cable and so must be removed fully before the cables can be withdrawn. Renew the sealing O-rings if damaged or worn. On refitting, spin the front wheel (or turn the engine over on the kickstart, as applicable) to help the drive mechanism engage the inner cable.

4   Check that the cables are routed correctly with no kinks or sharp bends, and that they are secured by any guide, clamps or ties provided for this purpose.

Speedometer – MTX125/200 RW-D models

5   The speedometer drive is fitted in the front brake backplate. The drive rarely gives trouble provided it is kept properly lubricated. Lubrication should take place whenever the front wheel is removed for wheel bearing or brake inspection or renewal, and consists of packing the recess with suitable grease.

6   Examine closely the large oil seal, renewing it if damaged or worn, and check its teeth for wear should the drive gear be removed at any time.

Speedometer – MBX125, MTX125 RW-F, H, L and MTX200 RW-F models

7   The speedometer drive gearbox is mounted on the front wheel, on the hub right-hand side on MTX125 RW-F, H, L and MTX200 RW-F models, and on the hub left-hand side on MBX125 models. The gearbox must be regarded as a sealed unit which cannot be repaired and requires no maintenance except for packing with grease whenever the wheel bearings are lubricated; note that the only component that can be removed if worn is the gearbox drive gear although the thrust washer(s) behind it are also listed separately.

8   The drive ring fitted in the hub can be removed as soon as the hub dust seal has been levered out. On refitting, ensure that the drive ring tabs are located correctly in the slots in the hub and that the dust seal is renewed. Smear a small amount of grease over the drive ring and seal lips before refitting the wheel to the machine. Note that the commonest fault with this type of speedometer drive is that the drive ring's raised ears are flattened by careless refitting of the gearbox; the ears can be bent back into position if care is exercised.

Tachometer – all models

9   The tachometer is driven by a worm gear from the oil pump drive shaft, the gear being housed in the crankcase right-hand cover. No maintenance is necessary, other than to check that all is in order whenever the crankcase cover is removed.

10   The gear is retained only by the fit of the oil seal at its upper end with the surrounding crankcase cover. If damaged or worn the gear can be pulled from its housing, or levered out from inside, noting the thrust washer fitted at each end. Always fit a new oil seal whenever it is disturbed in this way.

14.3 Drive cables are secured by knurled sleeve nuts at their upper ends

14.5 Speedometer drive is in front brake backplate on MTX125/200 RW-D models

14.9 Tachometer drive is mounted in crankcase right-hand cover

15 Bodywork: removal and refitting

1 The side panels, the fairing or headlamp nacelle, the radiator shroud(s) and the belly fairing (MBX125 only) are all lightweight plastic mouldings which must be removed and refitted with great care if they are not to be cracked or split.

2 On MBX125 models each side panel is mounted by three moulded prongs on its inner face engaging with rubber grommets set in the frame; these are removed by pulling carefully and are pressed into place on refitting. On MTX125/200 models the panels are retained in a similar fashion at two points, the remaining mounting being by a single screw set centrally in each panel; remove the screw and pull each panel outwards at the rear, then down at the front to release the mountings. Reverse the removal procedure on refitting.

3 The radiator shroud(s) and belly fairing (where fitted) are removed and refitted as described in Chapter 2. On MTX125/200 models take care not to break the bottom mounting hook of each shroud as it is removed or refitted. On MBX125 models, note that the radiator shroud and belly fairing are joined together and retained to the frame by small brackets extending from the radiator bottom mountings. If difficulty is encountered in removing or refitting the shroud or belly fairing, slacken the radiator bottom mounting bolts so that the brackets can be aligned correctly. Tighten all fasteners to the torque settings specified.

4 On MTX125/200 models, the headlamp nacelle is retained by two bolts, one on each side; do not forget to refit the spacers and tighten the bolts to a torque setting of 0.8 – 1.2 kgf m (6 – 9 lbf ft) on refitting.

5 On MBX125 models, the fairing is secured to the instrument panel mounting bracket by four bolts; two on each side projecting forwards just below the top yoke and two more facing sideways underneath the front of the fairing. Remove all four bolts, disconnect the turn signal lamp wires and withdraw the fairing. Check that the fairing is aligned correctly on its mountings and that a shouldered spacer is pressed correctly into each mounting rubber before tightening the mounting bolts.

15.5a Fairing removal MBX125 – remove upper rear mounting bolts (one each side) ...

15.5b ... followed by two lower front mounting bolts

Chapter 6 Wheels, brakes and tyres

Contents

Specifications

Unless otherwise stated, information applies to all models

Wheels	MBX125	MTX125/200
Size:		
Front	1.60 x 16	1.40 x 21
Rear	1.85 x 18	1.85 x 18
Rim maximum runout – radial and axial	2.0 mm (0.080 in)	
Spindle maximum runout	0.2 mm (0.008 in)	

Brakes	MBX125	MTX125 RW-F, H, L, MTX200 RW-F
Front disc brake:		
Disc standard thickness	3.80 - 4.20 mm (0.1496 - 0.1654 in)	3.5 mm (0.1377 in)
Service limit	3.00 mm (0.1181 in)	
Disc runout	0 - 0.15 mm (0 - 0.0059 in)	
Service limit	0.30 mm (0.0118 in)	
Master cylinder ID	12.700 - 12.743 mm (0.4999 - 0.5017 in)	
Service limit	12.755 mm (0.5022 in)	
Master cylinder piston OD	12.657 - 12.684 mm (0.4983 - 0.4994 in)	
Service limit	12.640 mm (0.4976 in)	
Caliper bore ID	25.400 - 25.405 mm (0.9999 - 1.0002 in)	
Service limit	25.450 mm (1.0020 in)	
Caliper piston OD	25.318 - 25.368 mm (0.9968 - 0.9987 in)	
Service limit	25.300 mm (0.9961 in)	

Drum brake – front or rear:
    Drum standard ID .... 110 mm (4.3307 in)
    Service limit .... 111 mm (4.3701 in)
    Friction material thickness .... 4 mm (0.16 in)
    Service limit .... 2 mm (0.08 in)

Tyres

	MBX125	MTX125/200
Size:		
Front	80/100 - 16 45P	2.75 x 21 4PR
Rear	90/90 - 18 51 P	4.10 x 18 4PR
Pressures – tyres cold:		
Front and rear	32 psi (2.25 kg/cm²	21 psi (1.50 kg/cm²)

Note: pressures are for solo or pillion use, except for MTX125/200 models where the rear tyre pressure should be increased to 25 psi (1.75 kg/cm²) when a pillion passenger is carried.

Manufacturer's recommended minimum tread depth – at centre of tread:

	MBX125	MTX125/200
Front	1.5 mm (0.06 in)	3.0 mm (0.12 in)
Rear	2.0 mm (0.08 in)	3.0 mm (0.12 in)

Torque wrench settings

Component	kgf m	lbf ft
Front wheel spindle nut		
MBX125	5.5 - 7.0	40 - 50.5
MTX125/200 RW-D	6.0 - 8.0	43 - 58
Front wheel spindle – MTX125 RW-F, H, L, MTX200 RW-F	5.0 - 8.0	36 - 58
Spindle clamp nuts – MTX125 RW-F, H, L, MTX200 RW-F	1.0 - 1.4	7 - 10
Brake disc mounting nuts – MBX125	2.7 - 3.3	19.5 - 24
Brake disc mounting bolts – MTX125 RW-F, H, L,		
MTX200 RW-F	1.4 – 1.6	10 – 12
Rear wheel spindle nut	6.0 - 8.0	43 - 58
Sprocket retaining nuts	5.5 - 6.5	40 - 47
Handlebar master cylinder clamp bolt:		
MBX125	1.0 - 1.4	7 - 10
MTX125 RW-F, H, L, MTX200 RW-F	0.8 – 1.2	6 – 9
Brake hose union bolts:		
MBX125	2.5 - 3.5	18 - 25
MTX125 RW-F, H, L, MTX200 RW-F	N/Av	N/Av
Brake hose clamp bolts – MBX125	2.0 - 2.5	14.5 - 18
Bleed nipple:		
MBX125	0.4 - 0.7	3 - 5
MTX125 RW-F, H, L, MTX200 RW-F	N/Av	N/Av
Caliper bracket/fork lower leg mounting bolts:		
MBX125	2.4 - 3.0	17 - 22
MTX125 RW-F. H, L, MTX200 RW-F	2.0 - 3.0	14.5 - 22
Pad pin bolts	1.5 - 2.0	11 – 14.5
Brake pedal pinch bolt:		
MBX125	0.8 - 1.2	6 - 9
MTX125/200	1.9 - 2.3	14 - 16.5
Drum brake operating arm pinch bolt	0.8 - 1.2	6 - 9

1 General description

On MBX125 models both wheels are of Honda's 'Comstar' type in which an aluminium alloy rim is attached to an alloy hub by three pairs of V-shaped spoke plates; these plates are riveted to the rim and bolted to the hub, but it is not intended that the wheels should be dismantled for any reason. On MTX125/200 models an aluminium alloy rim is laced to a light alloy hub by a number of steel wire spokes; although this method of construction does require regular checking as described in Routine Maintenance, it has the advantage that the wheels can be rebuilt, to repair accident damage for example, even if this is a task for the expert only.

On MBX125, MTX125 RW-F, H, L, and MTX200 RW-F models the front brake is hydraulically-operated; pressure applied via a handlebar-mounted master cylinder being transmitted to a twin piston caliper which acts on a disc bolted to the front hub. On MTX125/200 RW-D models a cable-operated drum brake is fitted; since only one operating camshaft is fitted, the brake is of the single-leading shoe type. The rear brake is a rod-operated single-leading shoe drum type on all models.

The tyres are of conventional type, with a separate inner tube.

This Chapter describes only the removal and refitting of the wheels, brakes and tyres and the overhaul of wheel bearings and brake components; for details of regular checks on the wheels, wheel bearings, brakes and tyres, for brake adjustment and, where applicable, for pad renewal, refer to Routine Maintenance.

2 Front wheel: removal and refitting

MBX125

1 Place the machine on its centre stand on level ground and remove the spindle nut, then remove its retaining screw from the drive gearbox and pull out the speedometer cable, Wedge wooden blocks or similar under the belly fairing, taking care not to scratch it, so that the machine is supported securely with the front wheel clear of the ground.

2 Tap out the wheel spindle and withdraw the wheel from the machine, then withdraw the hub right-hand spacer and the speedometer gearbox.

3 On refitting, insert the spacer into the hub right-hand side and the gearbox into its left-hand side, ensuring that the raised ears of the drive ring engage with the slots in the gearbox drive gear. Thoroughly clean

the spindle, removing all traces of dirt, corrosion and old grease, then smear clean grease along its length.

4   Offer up the wheel to the machine so that the disc passes between the brake pads and ensuring that neither the spacer nor the gearbox are displaced, then refit the spindle from right to left.

5   Rotate the speedometer gearbox so that its bottom edge is approximately horizontal; note that a lug on the gearbox should butt against a lug on the fork lower leg. Spin the wheel and insert the speedometer cable into its gearbox, tightening securely the retaining screw. Check that the cable is routed in a smooth curve, then tighten the spindle nut to a torque setting of 5.5 - 7.0 kgf m (40 - 50.5 lbf ft).

6   Apply the brake lever repeatedly until the pads are pushed back into contact with the disc and full brake pressure is restored. Check that the brake and speedometer are functioning correctly and that the wheel is free to rotate easily before taking the machine out on the road.

MTX125/200 RW-D

7   Remove its retaining screw and pull the speedometer cable out of the brake backplate. Slacken the two brake cable clamp bolts on the fork lower leg. Slacken fully its locknut and screw in fully the adjuster at the handlebar lever, then unscrew fully the (lower) adjuster locknut and displace the rubber cap so that the cable can be pulled upwards out of the adjuster abutment on the brake backplate. Slide the cable inner wire out through the slot provided and disengage the end nipple from the operating arm; do not lose the coil spring. Remove the spindle nut.

8   Lift the machine on to a stand so that it is supported securely upright with the front wheel clear of the ground, then tap out the spindle and withdraw the wheel from the machine. The hub right-hand spacer and the front brake backplate can then be withdrawn.

9   On refitting, insert the spacer into the hub right-hand side and the brake backplate into the drum, ensuring that the ears of the speedometer drivegear engage with the slots in the hub centre boss. Thoroughly clean the spindle, removing all traces of dirt, corrosion and old grease, then smear clean grease along its length.

10 Offer up the wheel to the machine ensuring that the groove in the brake backplate engages with the lug on the fork lower leg and that the spacer is not displaced, then refit the spindle from right to left. Fit the spindle nut, tightening it by hand only. Spin the wheel and insert the speedometer cable into its housing, tightening securely the retaining screw.

11 Reverse the dismantling procedure to refit the brake cable, then spin the wheel hard and apply the brake firmly to centralise the shoes and backplate on the drum. Maintain brake pressure while the spindle

nut is tightened to a torque setting of 6.0 - 8.0 kgf m (43 - 58 lbf ft). Adjust the brake as described in Routine Maintenance, not forgetting to tighten the cable clamp bolts securely. Check that the brake and speedometer are functioning correctly and that the wheel will rotate freely before taking the machine out on the road.

MTX125 RW-F, H, L and MTX200 RW-F

12 Slacken the four nuts securing the spindle clamp to the fork right-hand lower leg, then slacken the spindle itself; there is no need to remove the clamp completely.

13 Lift the machine on to a stand so that it is supported securely upright with the front wheel clear of the ground, then unscrew the spindle and withdraw it while holding the wheel in position. Noting carefully how it is aligned, withdraw the speedometer gearbox from the wheel and lower the wheel away from the machine. There is no need to disconnect the speedometer cable; the wheel can be rotated about the fork left-hand leg so that the gearbox can be detached easily. Once the wheel has been removed the hub left-hand spacer can be withdrawn.

14 On refitting, insert the spacer into the hub left-hand side and check that the thread in the fork left-hand lower leg and the spindle clamp on the right-hand lower leg are completely clean. If disturbed, the spindle clamp is refitted with the 'Up' mark facing upwards. Thoroughly clean the spindle, removing all traces of dirt, corrosion and old grease, then smear clean grease along its length.

15 Lift the wheel into position so that the disc passes between the brake pads and ensuring that the spacer is not displaced, then refit the speedometer gearbox ensuring that the raised ears of the drive ring engage with the slots in the gearbox drive gear. Align the wheel with both fork legs and refit the spindle, passing it through the loosely assembled clamp.

16 Tighten lightly the two upper clamp nuts then rotate the speedometer gearbox so that the cable runs straight to its clamp on the lower leg; a lug on the gearbox will butt against the lower leg to ensure this. Tighten the wheel spindle securely and apply the brake lever repeatedly until the pads are pushed back into contact with the disc and full brake pressure is restored.

17 Push the machine off its stand, apply the front brake and pump the forks vigorously up and down to align the right-hand lower leg on the spindle. Tighten securely first the upper two clamp nuts, then the lower two nuts; be careful not to overtighten them.

18 Before taking the machine out on the road, check that the forks, brake and speedometer are functioning correctly and that the wheel will rotate freely.

2.1a Remove wheel spindle nut ...

2.1b ... then unscrew retaining screw and withdraw speedometer cable

2.3a Insert spacer into hub right-hand side ...

2.3b ... and ensure raised ears of drive ring engage correctly on speedometer gearbox

2.5 Lug on gearbox should engage with lug on fork lower leg when cable is correctly routed

2.7 Disconnect brake cable from backplate, then release nipple from operating arm

2.9 Ears of speedometer drive gear must engage with slots in hub centre boss

2.10 Groove in brake backplate must engage with lug on fork lower leg

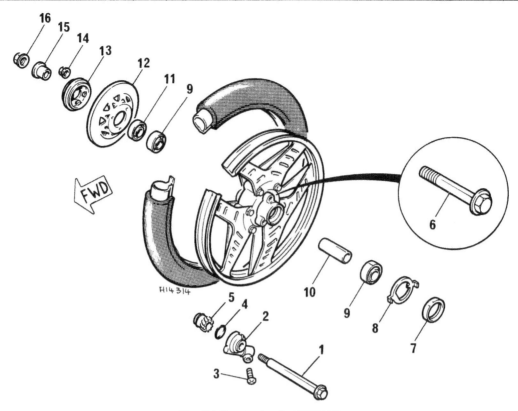

H14314

Fig. 6.1 Front wheel – MBX125

1	Spindle	5	Speedometer drive gear	9	Bearing – 2 off	13	Cover
2	Speedometer gearbox	6	Bolt – 3 off	10	Spacer	14	Nut – 3 off
3	Screw	7	Dust seal	11	Dust seal	15	Spacer
4	Thrust washer	8	Drive ring	12	Brake disc	16	Nut

H14315

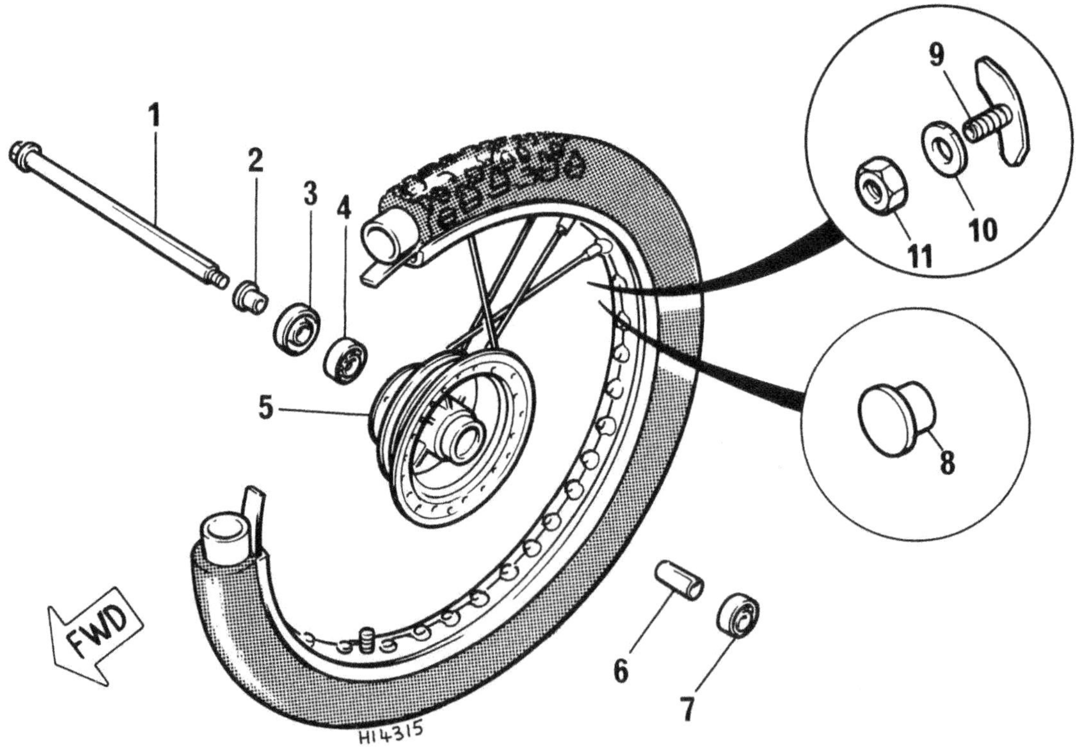

Fig. 6.2 Front wheel – MTX125 and 200 RW-D

1	Spindle	4	Right-hand bearing	7	Left-hand bearing	10	Washer*
2	Spacer	5	Wheel hub	8	Plug	11	Nut*
3	Dust seal	6	Spacer	9	Security bolt*		*Where fitted

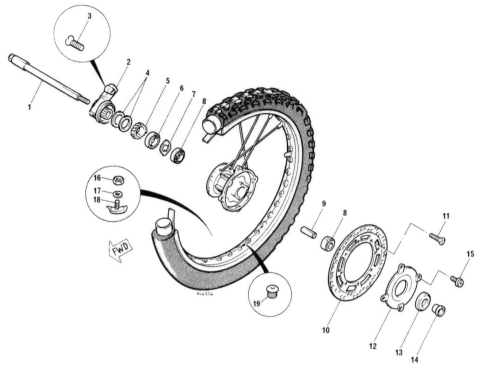

Fig. 6.3 Front wheel – MTX125 RW-F, H, L and MTX200 RW-F

1 Spindle	6 Dust seal	11 Bolt – 4 off	16 Nut
2 Speedometer gearbox	7 Drive ring	12 Cover	17 Washer
3 Screw	8 Bearing – 2 off	13 Oil seal	18 Security bolt
4 Thrust washer – 2 off	9 Spacer	14 Spacer	19 Plug
5 Speedometer gear	10 Brake disc	15 Screw – 4 off	

3 Rear wheel: removal and refitting

MBX125 and MTX200 RW-F models

1 On MBX125 models, place the machine securely on its centre stand on level ground and unscrew the spindle nut. On MTX200RW-F models, slacken the spindle nut, then lift the machine on to a stand so that it is supported securely upright with the rear wheel clear of the ground before unscrewing the nut.

2 Unscrew the brake adjusting nut and disconnect the operating rod from the arm on the brake backplate, then tap out the wheel spindle. Disengage the chain from the sprocket and hang it over the swinging arm fork end, then withdraw the wheel from the machine. The hub left-hand spacer and brake backplate may then be withdrawn.

3 On refitting, check that the chain adjusters are completely clean, well greased and that both are inserted correctly into the swinging arm fork ends so that the chain adjustment index marks are visible from the outside. Insert the spacer into the hub left-hand side and the brake backplate into the drum. Thoroughly clean the spindle, removing all traces of dirt, old grease and corrosion, then smear clean grease along its length.

4 Offer up the wheel to the machine, ensuring that the groove on the brake backplate engages with the lug on the inside of the swinging arm right-hand fork end, and that neither the hub left-hand spacer or the washer crimped on the backplate boss are displaced. Engage the chain on the sprocket and refit the spindle from right to left. Refit the spindle nut, tightening it by hand only at this stage and connect the brake rod to the operating arm, not forgetting the coil spring.

5 Adjust the drive chain tension as described in Routine Maintenance, then adjust the brake and stop lamp switch. Before using the machine on the road, check that all disturbed components are securely fastened, that the chain is correctly adjusted and lubricated, that the wheel is free to rotate easily and that the brake and stop lamp switch are functioning correctly.

MTX125 and MTX200 RW-D models

6 Support the machine securely on a stand so that the rear wheel is raised clear of the ground, then unscrew the brake adjusting nut and disconnect the brake rod from the operating arm.

7 Slacken the spindle nut sufficiently to enable both snail cams to be rotated forwards until they can be hooked over the stopper pins, disengage the chain from the sprocket and hang it over the swinging arm fork end, then slacken the spindle nut further until the stopper plate on the right-hand side can be unhooked from the stopper pin. The wheel assembly can then be removed from the machine. Unscrew the spindle nut and withdraw the washer, the left-hand snail cam and the hub left-hand spacer, then tap out the spindle and withdraw the brake backplate, the stopper-plate and the right-hand snail cam.

8 On reassembly, thoroughly clean the spindle, removing all traces of dirt, corrosion and old grease, then smear clean grease over the length of it and refit it from right to left, passing through the right-hand snail cam, the stopper plate, the brake backplate, the wheel hub, the hub left-hand spacer and the left-hand snail cam before the spindle nut is refitted. Note that the snail cams are fitted with their marked surfaces outwards and that they are handed; the letter 'L' or 'R' stamped in each cam will identify it as a left-hand or right-hand component.

9 Offer the wheel assembly up to the machine ensuring that the lug on the swinging arm right-hand fork inside end engages correctly with the groove in the brake backplate and that the small hole in the stopper plate fits over the stopper pin. Hook the snail cams over the stopper pins and tighten the wheel spindle nut until the cams are just free to move, then engage the chain on the sprocket.

10 Adjust the drive chain tension as described in Routine Maintenance, then adjust the brake and stop lamp switch. Check that the chain is correctly adjusted and lubricated, that the rear wheel is free to revolve smoothly and easily, that the rear brake and stop lamp switch are adjusted correctly and operating properly and that all disturbed components are securely fastened.

All models

11 If the brake appears spongy after chain adjustment or rear wheel removal, slacken the spindle nut and apply the brake firmly to centralise the shoes and backplate on the drum. Maintain brake pressure while tightening the spindle nut to the specified torque setting.

3.3 Chain adjusters must be clean and greased, with alignment marks visible on outside

3.4a Do not displace hub left-hand spacer on refitting wheel ...

3.4b ... and be careful not to lose washer crimped on to brake backplate boss

3.8a Do not omit stopper plate on refitting wheel ...

3.8b ... and note that snail cams are handed – refit the correct way round

3.9 Groove in brake backplate must engage with swinging arm lug

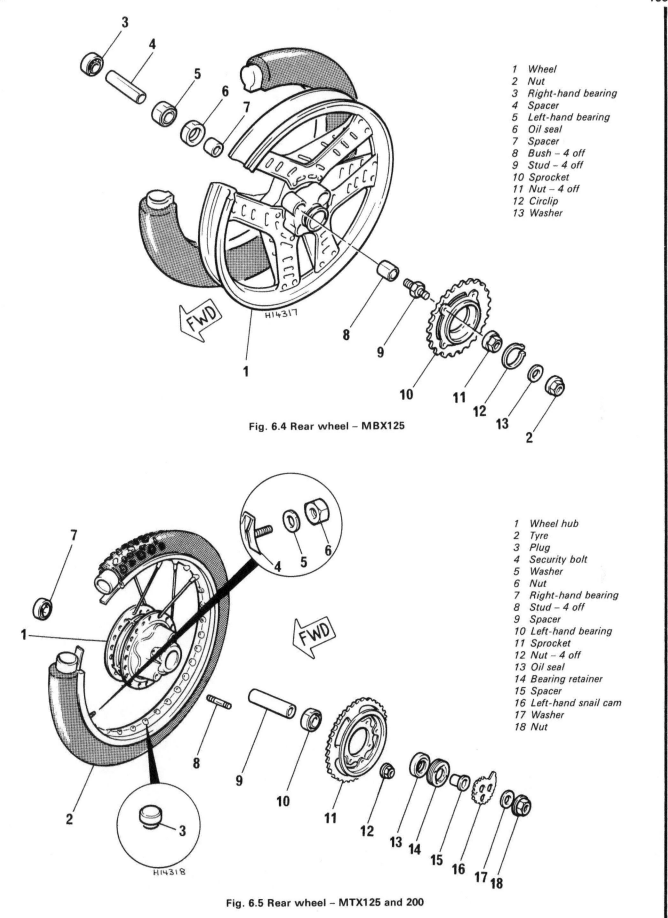

1 Wheel
2 Nut
3 Right-hand bearing
4 Spacer
5 Left-hand bearing
6 Oil seal
7 Spacer
8 Bush – 4 off
9 Stud – 4 off
10 Sprocket
11 Nut – 4 off
12 Circlip
13 Washer

Fig. 6.4 Rear wheel – MBX125

1 Wheel hub
2 Tyre
3 Plug
4 Security bolt
5 Washer
6 Nut
7 Right-hand bearing
8 Stud – 4 off
9 Spacer
10 Left-hand bearing
11 Sprocket
12 Nut – 4 off
13 Oil seal
14 Bearing retainer
15 Spacer
16 Left-hand snail cam
17 Washer
18 Nut

Fig. 6.5 Rear wheel – MTX125 and 200

4   Wheel bearings: removal, examination and refitting

1   Remove from the machine the wheel to be overhauled and withdraw the hub spacer and, where applicable, the brake backplate; refer to Section 2 or 3 of this Chapter.

2   On the front wheel of MBX125 machines, carefully lever off the plastic cap from the centre of the brake disc. On all machines with disc front brakes, the speedometer drive ring can be lifted out of the hub after the retaining oil seal on that side has been levered from its housing.

3   On the rear wheel of MTX125/200 machines, the hub left-hand wheel bearing and oil seal are secured by a threaded retainer which is screwed into the hub and staked at four points to prevent it from unscrewing. The retainer is removed and refitted using a special tool. Part Number 07710-0010401 in conjunction with an adaptor, Part Number 07710-0010600. If these are not available, and a peg spanner of suitable size cannot be found, obtain a steep strip (an old tyre lever would be ideal) and drill two holes in it to correspond with two diagonally opposite holes in the retainer. Pass a small bolt through each hole and secure them with nuts to complete a fabricated peg spanner of the type shown in the accompanying illustration.

4   In all cases, the oil seal into which the hub spacer is inserted may be either levered out or driven out with the appropriate bearing.

5   The bearings are easiest to remove if an internally-expanding bearing puller is used. If this is not available an expanding bolt such as a Rawlbolt could be tightened on to the inner race of one bearing so that it can be driven out by passing a drift through the hub. The simplest method is given below but will require some skill and care if it is to be successful. On MTX125/200 machines the rear wheel left-hand bearing must be removed first, but in all other cases the right-hand bearing must be removed first; locating ribs inside the hub mean that the central spacer can be displaced at one end only.

6   Position the wheel on a work surface with its hub well supported by wooden blocks so that enough clearance is left beneath the wheel to drive out the first bearing. Ensure the blocks are placed as close to the bearing as possible, to lessen the risk of distortion of the hub casting whilst the bearings are being removed or refitted.

7   Place the end of a long-handled drift through the hub and against the upper face of the bearing inner race and tap the bearing downwards out of the wheel hub. The spacer located between the two bearings may be moved sideways slightly in order to allow the drift to be positioned against the face of the bearing. Move the drift around the face of the bearing whilst drifting it out of position, so that the bearing leaves the hub squarely.

8   With the one bearing removed, the wheel may be lifted and the spacer withdrawn from the hub. Invert the wheel and remove the

second bearing, using a similar procedure. The dust seal should be closely inspected for any indication of damage, hardening or perishing and renewed if necessary. It is advisable to renew all seals as a matter of course if the bearings are found to be defective.

9   Remove all the old grease from the hub and bearings, giving the latter a final wash in petrol. Check the bearings for signs of play or roughness when they are turned. If there is any doubt about the condition of a bearing, it should be renewed.

10   If the original bearings are to be refitted, they should be repacked with high-melting grease before being fitted into the hub. New bearings must also be packed with grease. Ensure that the bearing recesses in the hub are clean and both bearings and recess mating surfaces lightly greased to aid fitting. Check the condition of the hub recesses for evidence of abnormal wear which may have been caused by the outer race of a bearing spinning. If evidence of this is found, and the bearing is a loose fit in the hub, it is best to seek advice from a Honda Service Agent or a competent motorcycle engineer. Alternatively, a proprietary product such as Loctite Bearing Fit may be used to retain the bearing outer race; this will mean, however, that the bearing housing must be cleaned and degreased before the locking compound can be used.

11   With the wheel hub and bearing thus prepared, fit the bearings and central spacer as follows. With the hub again well supported by the wooden blocks, drift the first of two bearings into position. Use a soft-faced hammer in conjunction with a socket or length of metal tube which has an overall diameter which is slightly less than that of the outer race of the bearing, but which does not bear at any point on the bearing sealed surface or inner race. Tap the bearing into place against the locating shoulder machined in the hub, remembering that the sealed surface of the bearing must always face outwards. With the first bearing in place, invert the wheel, insert the central spacer and pack the hub centre no more than 2/3 full with high-melting point grease. Fit the second bearing, using the same procedure. Take great care to ensure that each of the bearings enters its housing correctly, that is, square to the housing, otherwise the housing surface may be broached.

12   On machines with disc front brakes, insert the speedometer drive ring into the hub so that its protruding tangs fit into the slots in the hub. Oil seals should be greased before being drifted into place using the method described above. On the rear wheel of MTX125/200 machines, renew the retainer if its threads are damaged, apply grease to the threads and screw it into the hub until it is securely tightened. Use a hammer and a punch to stake it to the hub at four diametrically opposite points around its outer edge to prevent any risk of its slackening. On the front wheel of MBX125 machines tap the plastic cap into the hub right-hand side.

4.2a Plastic cap should be carefully prised off – MBX125

4.2b On machines with disc front brakes lever oil seal out of the hub ...

4.2c ... and withdraw the speedometer drive ring

4.3a MTX125/200 – peg spanner can be fabricated ...

4.3b ... to unscrew bearing retainer

4.7 Pass long drift through hub to tap out opposite bearing

4.11a Fit first bearing then invert wheel and fit central spacer – grease cavity ...

4.11b ... and fit second bearing ...

4.11c ... using a hammer and a tubular drift

4.12a Grease seals to aid refitting

4.12b Do not omit speedometer drive ring (where fitted)

4.12c MTX125/200 – tighten retainer securely then stake to prevent slackening

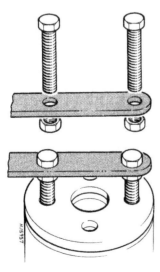

Fig. 6.6 Fabricated peg spanner for removing rear wheel bearing retainer

5 Rear sprocket and cush drive: removal, examination and refitting

1   On MTX125/200 models the sprocket is mounted on the hub by four nuts which are threaded on to studs screwed into the hub. Check at regular intervals (see Routine Maintenance) that the sprocket teeth are not hooked, chipped or badly worn and that the mounting nuts are securely fastened. If the sprocket is allowed to work loose, severe damage may be caused to the hub. No form of cush drive is fitted.

2   On MBX125 models four pins are secured to the sprocket by retaining nuts and engage with individual bonded rubber bushes pressed into the hub; the sprocket is retained by a large circlip. The rubber bushes permit a small amount of movement between the hub and sprocket to help damp out transmission shock loads; this shock absorber, or cush drive, must be checked at regular intervals (see Routine Maintenance) to ensure that the rubbers are not worn, producing excessive free movement in the normal direction of rotation, and that the retaining circlip is securely seated in its groove. Check also that the sprocket teeth are not hooked, chipped or badly worn.

3   If any sign of wear or damage is found, the rear wheel must be removed from the machine as described in Section 3 of this Chapter.

4   On MTX125/200 models remove the four nuts and lift away the

sprocket. Renew the sprocket if necessary and check that there is no free play when it is refitted over the studs. If any stud is worn, permitting some movement in the normal direction of rotation, the studs should be renewed as a set. Lock two nuts together on the exposed thread and apply an open-ended spanner to the lower nut to unscrew each stud. On refitting use the same technique but applied to the upper nut; apply a few drops of thread locking compound to each stud as it is screwed into the hub. On refitting the sprocket apply thread locking compound to the stud outer threads and thoroughly clean the hub and sprocket mating surfaces. Tighten the retaining nuts to a torque setting of 5.5 - 6.5 kgf m (40 - 47 lbf ft).

5 On MBX125 models, remove the large circlip to release the sprocket, which can then be lifted away. If this proves difficult due to the four pins being corroded into their bushes, unscrew the retaining nuts and lift the sprocket off the pins. If the sprocket is to be renewed unscrew the four nuts and withdraw the pins. On refitting, note that the flats on each pin must engage with the groove in the sprocket rear face; apply thread locking compound to the pin threads before refitting the nuts and tightening them to a torque setting of 5.5 - 6.5 kgf m (40 - 47 lbf ft). Thoroughly clean the hub and sprocket mating surfaces and the cush drive bush centres and sprocket pins, then smear high-melting point grease over the pins to prevent corrosion before refitting the sprocket. Check that the retaining circlip is securely seated in its groove.

6 The cush drive bushes must be renewed if they are cracked, perished or worn in any way which permits excessive sprocket movement. First remove the sprocket, as described above, then apply a liberal quantity of penetrating fluid and leave the wheel for some time, preferably at least overnight, so that the fluid can work; bonded rubber bushes, when pressed into blind holes as in this case, are almost impossible to remove due to the effects of corrosion and the extremely tight fit of the bush metal outer sleeves in the hub.

7 If the sprocket pins are still stuck in the bushes after the penetrating fluid has had time to work, refit the sprocket temporarily and use a legged sprocket puller to extract it, complete with the pins. It is essential that a thick metal spacer of suitable size is placed over the hub centre boss so that the puller centre bolt applies pressure directly to the hub itself, rather than to the bearings.

8 Some means must now be devised of extracting the bushes without damaging the hub. Do not heat the hub or apply excessive force to it; if the hub is damaged or distorted, the complete rear wheel must be renewed. Since any attempt at removal is likely to tear out the bush inner sleeve, thus making removal even more difficult, it is recommended that the wheel be cleaned, treated with penetrating fluid and taken to a Honda Service Agent for the bushes to be extracted and new ones fitted.

9 If the bushes are to be refitted at home, thoroughly clean the hub passages and use emery paper to polish away any burrs from the hub and from the bush outer sleeve. Smear grease over the bush to aid the task and tap each bush into the hub until its metal outer sleeve end is flush with the hub. Use a hammer and a tubular drift such as a socket spanner which bears only on the bush outer sleeve.

---

6 Front brake disc: examination, removal and refitting

---

1 The brake disc can be checked for wear and for warpage whilst the front wheel is still in the machine. Using a micrometer, measure the thickness of the disc at the point of greatest wear. If the measurement is much less than the recommended service limit the disc should be renewed. Check the warpage (runout) of the disc by setting up a suitable pointer close to the outer periphery of the disc and spinning the front wheel slowly. If the total warpage is more than 0.30 mm (0.012 in), the disc should be renewed. A warped disc, apart from reducing the braking efficiency, is likely to cause juddering during braking and will cause the brake to bind when it is not in use.

2 The brake disc should also be checked for bad scoring on its contact area with the brake pads. If any of the above mentioned faults are found, then the disc should be removed from the wheel for renewal or for repair by skimming. Such repairs should be entrusted only to a reputable engineering firm. A local motorcycle dealer may be able to assist in having the work carried out.

3 Remove the front wheel from the machine as described in Section 2 of this Chapter. On MBX125 models carefully lever off the plastic cap

from the centre of the disc, unscrew the three retaining nuts and withdraw the disc. On MTX125/200 models unscrew the four screws and lift away the cover from the centre of the disc, then unscrew the four mounting bolts and withdraw the disc.

4 Refitting is the reverse of the above, but when tightening the retaining nuts do so in an even and diagonal sequence to avoid stress on the disc. Secure to the recommended torque setting.

5.2a MBX125 – remove large circlip ...

5.2b ... and withdraw sprocket with cush drive pins

---

7 Master cylinder: examination and renovation

---

1 If the regular checks described in Routine Maintenance reveal the presence of a fluid leak, or if a loss of brake pressure is encountered at any time, the handlebar-mounted master cylinder assembly must be removed and dismantled for checking.

2 Disconnect the stop lamp switch wires at the switch, or behind the headlamp, as applicable. The switch need not be disturbed unless the master cylinder is to be renewed. Place a clean container beneath the caliper unit and run a clear plastic tube from the caliper bleed nipple to

the container. Unscrew the bleed nipple by one full turn and drain the system by operating the brake lever repeatedly until no more fluid can be seen issuing from the nipple.

3   Position a pad of clean cloth rag beneath the point where the brake hose joins the master cylinder to prevent brake fluid from dripping. Detach the rubber cover from the head of the union bolt and remove the bolt (MBX125 models); on MTX125/200 models unscrew the union gland nuts, then unscrew the union itself. Once any excess fluid has drained from the union connection, wrap the end of the hose in rag or polythene and then attach it to a point on the handlebars.

4   Remove the brake lever by unscrewing its shouldered pivot bolt and locknut, remove the reservoir cover and diaphragm, then remove the two clamp bolts and withdraw the master cylinder assembly.

5   Use the flat of a small screwdriver to prise out the rubber dust seal boot. This will expose a retaining circlip which must be removed using a pair of circlip pliers which have long, straight jaws. With the circlip removed, the piston and cup assembly can be pulled out. Be very careful to note the exact order in which these components are fitted.

6   Note that if a vice is used to hold the master cylinder at any time during dismantling and reassembly, its jaws must be padded with soft alloy or wooden covers and the master cylinder must be wrapped in soft cloth to obviate any risk of the assembly being marked or distorted.

7   Place all the master cylinder component parts in a clean container and wash each part thoroughly in new brake fluid. Lay the parts out on a sheet of clean paper and examine each one as follows.

8   Inspect the unit body for signs of stress failure around both the brake lever pivot lugs and the handlebar mounting points. Carry out a similar inspection around the hose union boss. Examine the cylinder bore for signs of scoring or pitting. If any of these faults are found, then the unit body must be renewed.

9   Inspect the surface of the piston for signs of scoring or pitting and renew it if necessary. It is advisable to discard all the components of the piston assembly as a matter of course as the replacement cost is relatively small and does not warrant re-use of components vital to safety. If measuring equipment is available, compare the dimensions of

the master cylinder bore and the piston with those given in the Specifications Section of this Chapter and renew any component that is worn beyond the set wear limit. Inspect the threads of the brake hose union (bolt) for any signs of failure and renew the bolt if in the slightest doubt. Renew each of the gasket washers located one either side of the hose union.

10 Check before reassembly that any traces of contamination remaining within the reservoir body have been removed. Inspect the diaphragm to see that it is not perished or split. It must be noted at this point that any reassembly work must be undertaken in ultra-clean conditions. Particles of dirt entering the component will only serve to score the working points of the cylinder and thereby cause early failure of the system.

11 When reassembling and fitting the master cylinder, follow the removal and dismantling procedures in the reverse order whilst paying particular attention to the following points. Make sure that the piston components are fitted the correct way round and in the correct order. Immerse all of these components in new brake fluid prior to reassembly and refer to the figure accompanying this text when in doubt as to their fitted positions. When refitting the master cylinder assembly to the handlebar, position the assembly so that the reservoir will be exactly horizontal when the machine is in use. Tighten the clamp top bolt first, and then the bottom bolt, to a torque setting of 0.8 - 1.2 kgf m (6 - 9 lbf ft). Connect the brake hose to the master cylinder, ensuring that a new sealing washer is placed on each side of the hose union, and tightening the hose union (bolt) to the specified torque setting. Finally, replace the rubber union cover (MBX125 models only).

12 Bleed the brake system after refilling the reservoir with new hydraulic fluid, then check for leakage of fluid whilst applying the brake lever. Push the machine forward and bring it to a halt by applying the brake. Do this several times to ensure that the brake is operating correctly before taking the machine for a test run. During the run, use the brakes as often as possible and on completion, recheck for signs of fluid loss.

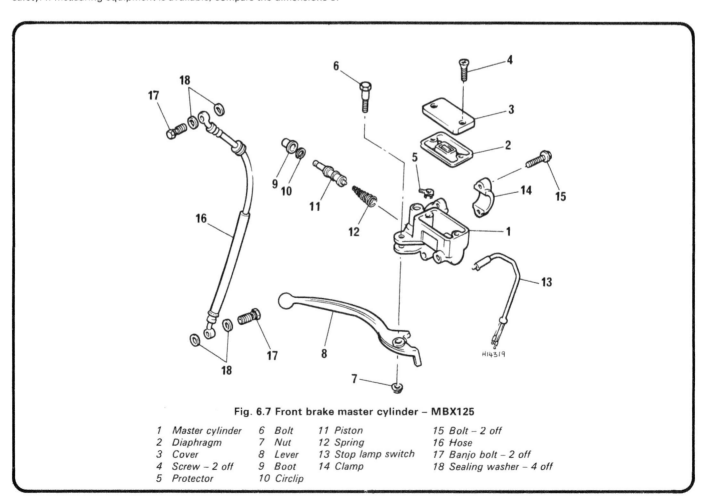

Fig. 6.7 Front brake master cylinder – MBX125

1 Master cylinder	6 Bolt	11 Piston	15 Bolt – 2 off
2 Diaphragm	7 Nut	12 Spring	16 Hose
3 Cover	8 Lever	13 Stop lamp switch	17 Banjo bolt – 2 off
4 Screw – 2 off	9 Boot	14 Clamp	18 Sealing washer – 4 off
5 Protector	10 Circlip		

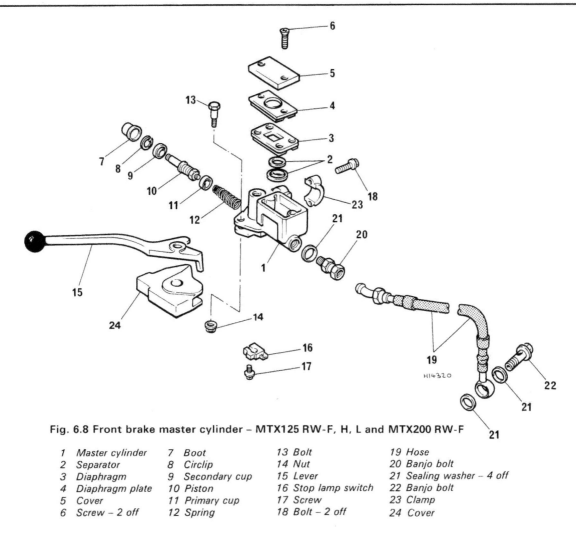

Fig. 6.8 Front brake master cylinder – MTX125 RW-F, H, L and MTX200 RW-F

1	Master cylinder	7	Boot	13	Bolt
2	Separator	8	Circlip	14	Nut
3	Diaphragm	9	Secondary cup	15	Lever
4	Diaphragm plate	10	Piston	16	Stop lamp switch
5	Cover	11	Primary cup	17	Screw
6	Screw – 2 off	12	Spring	18	Bolt – 2 off

19	Hose
20	Banjo bolt
21	Sealing washer – 4 off
22	Banjo bolt
23	Clamp
24	Cover

8 Brake hose: examination and renovation

1   A flexible brake hose is used as a means of transmitting hydraulic pressure to the caliper unit once the front brake lever is applied.

2   When the brake assembly is being overhauled, or at any time during a routine maintenance or cleaning procedure, check the condition of the hose for signs of leakage, damage, deterioration or scuffing against any cycle components. Any such damage will mean that the hose must be renewed immediately. The union connections at either end of the hose must also be in good condition, with no stripped threads or damaged sealing washers. Do not tighten these union bolts over the recommended torque setting as they are easily sheared if overtightened.

9 Brake caliper: examination and renovation

1   If the regular checks described in Routine Maintenance reveal the presence of any fluid leaks or of any wear, damage, or corrosion which will impair the caliper's efficiency, the unit must be removed and dismantled for checking. Start by removing the brake pads as described in Routine Maintenance.

2   Pull the mounting bracket out of the caliper body. If they are damaged or worn, the two axle bolts can be unscrewed from the bracket and renewed individually. Detach the rubber dust seal from the upper axle bolt and pull out of the caliper body the rubber dust seal/bush which fits around the lower axle bolt; these should be renewed if cracked, split or worn in any way.

3   The simplest way of ejecting the pistons from the caliper bores is to place the caliper in a plastic bag, tying the neck tightly around the brake hose to catch the inevitable shower of brake fluid. Apply the brake lever repeatedly, using normal hydraulic pressure to force out the piston. Ensure that the pistons leave the bores at the same time; if one sticks at any point the other piston must be restrained by firm hand pressure so that the full hydraulic pressure can overcome the resistance. It would be very difficult to extract one piston alone from this type of caliper without risking damage. When the pistons have been ejected, apply the brake lever until all remaining fluid is pumped into the bag, then unscrew the union bolt to disconnect the brake hose and remove the caliper from the machine, taking care to prevent the spillage of any brake fluid.

4   An alternative to the above method will require a source of compressed air. Attach a length of clear plastic tubing to the bleed nipple, placing the tube lower end in a suitable container. Open the bleed nipple by one full turn and pump gently on the front brake lever to drain as much fluid as possible from the hydraulic system. When no more fluid can be seen issuing from the bleed nipple, tighten it down again, withdraw the plastic tube and disconnect the hydraulic hose at the caliper union. Wrap a large piece of cloth loosely around the caliper and apply a jet of compressed air to the union orifice. Be careful not to use too high an air pressure or the pistons may be damaged.

5   If either of the above methods fails to dislodge the pistons, no further time should be wasted and the complete caliper assembly should be renewed. If the pistons and caliper bores are so badly damaged or so corroded that hydraulic or air pressure cannot move the pistons, the caliper will not be safe for further use even if a method can be devised of removing the pistons and cleaning the bearing surfaces.

6   When each piston has been removed, it should be placed in a clean container to ensure that it cannot be damaged and the seals should be picked out of the caliper bores, pressing each inwards to release it, and

discarded. Wrap the exposed end of the hydraulic hose in clean rag or polythene to prevent the entry of dirt. Carefully wash off all spilt brake fluid from the machine using fresh water; brake fluid is an excellent paint stripper and will attack any painted metal or plastic component.

7   Clean the caliper components thoroughly in hydraulic fluid. **Never** use petrol or cleaning solvent for cleaning hydraulic brake parts otherwise the rubber components will be damaged. Discard all the rubber components as a matter of course. The replacement cost is relatively small and does not warrant re-use of components vital to safety. Check the pistons and caliper cylinder bores for scoring, rusting or pitting. If any of these defects are evident it is unlikely that a good fluid seal can be maintained and for this reason the components should be renewed. If measuring equipment is available, compare the dimensions of the caliper bores and the pistons with those given in the Specifications Section of this Chapter, renewing any component that is worn to beyond the set wear limit.

8   Inspect the shank of each axle bolt for any signs of damage or corrosion and clean or renew each one as necessary; if the matching bores in the caliper body are worn or damaged, the body must be renewed. It is essential that the body can move smoothly and easily on the axle bolts. Check also that the bleed nipple is clear and its threads undamaged, and that the dust cap is in good condition; renew either component if faulty.

9   On reassembly, the components must be absolutely clean and dry; cleanliness is essential to prevent the entry of dirt into the system. Soak the new seals in hydraulic fluid before refitting them to the caliper

bores; check that each seal is securely located in its groove without being twisted or distorted. Smear a liberal quantity of clean brake fluid over the pistons and caliper bores then very carefully insert the pistons (with their recessed faces outwards) with a twisting motion, as if screwing them in, to avoid damaging or dislodging the seals. When the piston outer ends project 3-5 mm above the caliper they are fully refitted; do not press them fully into the caliper bores. Wipe off any surplus brake fluid.

10 Fit the rubber bush and seal to the caliper body and mounting bracket, check that the axle bolts are tightened securely on the bracket and smear silicone- or PBC-based brake caliper grease over the bolt bearing surfaces. Pack a small amount of grease into the matching caliper passages and refit the mounting bracket, ensuring that the seals are correctly engaged on their locating shoulders. Renew the sealing washers if worn or distorted and connect the brake hose to the caliper. Check that the hose is correctly aligned and tighten the union bolt to the specified torque setting. Refit the pads to the caliper and the caliper assembly to the machine, as described in Routine Maintenance.

11 Refill the master cylinder reservoir with new hydraulic fluid and bleed the system by following the procedure given in Section 10 of this Chapter. On completion of bleeding, carry out a check for leakage of fluid whilst applying the brake lever. Push the machine forward and bring it to a halt by applying the brake. Do this several times to ensure that the brake is operating correctly before taking the machine out on the road.

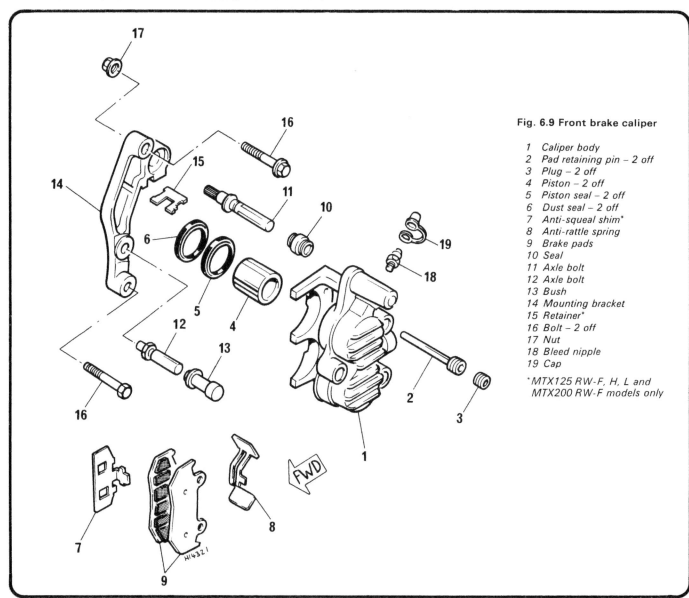

Fig. 6.9 Front brake caliper

1   Caliper body
2   Pad retaining pin – 2 off
3   Plug – 2 off
4   Piston – 2 off
5   Piston seal – 2 off
6   Dust seal – 2 off
7   Anti-squeal shim*
8   Anti-rattle spring
9   Brake pads
10 Seal
11 Axle bolt
12 Axle bolt
13 Bush
14 Mounting bracket
15 Retainer*
16 Bolt – 2 off
17 Nut
18 Bleed nipple
19 Cap

*MTX125 RW-F, H, L and
MTX200 RW-F models only

10 Bleeding the hydraulic brake system

1 If the brake action becomes spongy, or if any part of the hydraulic system is dismantled it is necessary to bleed the system in order to remove all traces of air. The procedure for bleeding the hydraulic system is best carried out by two people.

2 Check the fluid level in the reservoir and top up with fluid of the specified type if required. Keep the reservoir at last half full during the bleeding procedure; if the level is allowed to fall too far, air will enter the system requiring that the procedure be started again from scratch. Refit the reservoir cover to prevent the ingress of dust or the ejection of a spout of fluid.

3 Remove the dust cap from the caliper bleed nipple and clean the area with a rag. Place a clean glass jar below the caliper and connect a pipe from the bleed nipple to the jar. A clear plastic tube should be used so that air bubbles can be more easily seen. Place some clean hydraulic fluid in the glass jar so that the pipe is immersed below the fluid surface throughout the operation.

4 If parts of the system have been renewed, and thus the system must be filled, open the bleed nipple about one turn and pump the brake lever until fluid starts to issue from the clear tube. Tighten the bleed nipple and then continue the normal bleeding operation as described in the following paragraphs. Keep a close check on the reservoir level whilst the system is being filled.

5 Operate the brake lever as far as it will go and hold it in this position against the fluid pressure. If spongy brake operation has occurred it may be necessary to pump the brake lever rapidly a number of times until pressure is achieved. With pressure applied, loosen the bleed nipple about half a turn. Tighten the nipple as soon as the lever has reached its full travel and then release the lever. Repeat this operation until no more air bubbles are expelled with the fluid into the glass jar. When this condition is reached, the air bleeding operation should be complete, resulting in a firm feel to the brake operation. If sponginess is still evident continue the bleeding operation; it may be that an air bubble trapped at the top of the system has yet to work down through the caliper.

6 When all traces of air have been removed from the system, top up the reservoir and refit the diaphragm and cover. Check the entire system for leaks, and check also that the brake system in general is functioning efficiently before using the machine on the road.

7 Brake fluid drained from the system will almost certainly be contaminated, either by foreign matter or more commonly by the absorption of water from the air. All hydraulic fluids are to some degree hygroscopic, that is, they are capable of drawing water from the atmosphere, and thereby degrading their specifications. In view of this, and the relative cheapness of the fluid, old fluid should always be discarded.

8 Great care should be taken not to spill hydraulic fluid on any painted cycle parts; it is very effective paint stripper. Also, the plastic glasses in the instrument heads, and most other plastic parts, will be damaged by contact with this fluid.

11 Drum brakes: overhaul – all models

1 The drum brake fitted to the front wheel of MTX125/200RW-D models is the same in design and construction as that fitted to the rear wheel of all models. If the regular checks described in Routine Maintenance reveal the brakes to be faulty or the friction material to be worn out, the relevant wheel must be removed from the machine as described in Section 2 or 3 of this Chapter. Note also that the brakes must be overhauled, as described below, as a part of the maintenance schedule whether they are faulty or not.

2 Measure the thickness of friction material at several points on each shoe; if the material is worn at any point to less than the specified limit, the brake shoes should be renewed. The linings are bonded on and cannot be supplied separately.

3 If oil or grease from the wheel bearings has badly contaminated the linings, the brake shoes should be renewed; there is no satisfactory way of degreasing the lining material. Any surface dirt on the linings can be removed with a stiff-bristled brush and high spots on the linings should be carefully eased down with emery cloth.

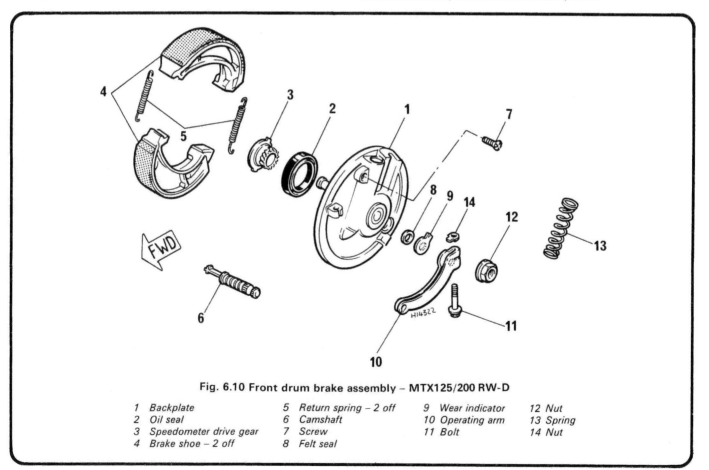

Fig. 6.10 Front drum brake assembly – MTX125/200 RW-D

1 Backplate	5 Return spring – 2 off	9 Wear indicator	12 Nut
2 Oil seal	6 Camshaft	10 Operating arm	13 Spring
3 Speedometer drive gear	7 Screw	11 Bolt	14 Nut
4 Brake shoe – 2 off	8 Felt seal		

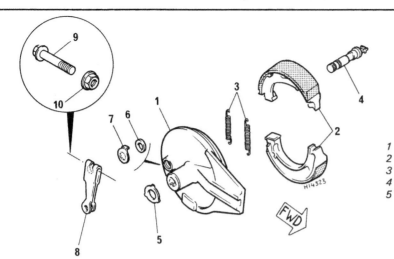

Fig. 6.11 Rear brake assembly

1 Backplate	6 Felt seal
2 Brake shoe – 2 off	7 Wear indicator
3 Return spring – 2 off	8 Operating arm
4 Camshaft	9 Bolt
5 Washer	10 Nut

4   Examine the drum surface for signs of scoring, wear beyond the service limit or oil contamination. All of these conditions will impair braking efficiency. Remove all traces of dust, preferably using a brass wire brush, taking care not to inhale any of it as it is of an asbestos nature, and consequently harmful. Remove oil or grease deposits, using a petrol soaked rag.

5   If deep scoring is evident, due to the linings having worn through to the shoe at some time, the drum must be skimmed on a lathe, provided that this does not exceed the service limit, or renewed. Whilst there are firms who will skim the drum without dismantling the wheel, it should be borne in mind that excessive skimming will change the radius of the drum in relation to the brake shoes, thereby reducing the friction area until extensive bedding in has taken place. Also full adjustment of the shoes may not be possible. If in doubt about this point, the advice of one of the specialist engineering firms who undertake this work should be sought.

6   It is a false economy to try to cut corners with brake components, the whole safety of both machine and rider being dependent on their good condition.

7   Removal of the brake shoes is accomplished by folding the shoes together so that they form a 'V'. With the spring tension relaxed, both shoes and springs may be removed from the brake backplate as an assembly. Detach the springs from the shoes and carefully inspect them for any signs of fatigue or failure. If in doubt, compare them with a new set of springs.

8   Before fitting the brake shoes, check that the brake operating cam is working smoothly and is not binding in its pivot. The cam can be removed by withdrawing the pinch bolt on the operating arm and pulling the arm off the shaft. Before removing the arm, if the manufacturer's alignment punch marks cannot be seen, it is advisable

to mark its position in relation to the shaft so that it can be relocated correctly with the wear indicator pointer.

9   Remove any deposits of hardened grease or corrosion from the bearing surface of the brake cam and shoe by rubbing it lightly with a strip of fine emery paper or by applying solvent with a piece of rag. Lightly grease the length of the shaft and the face of the operating cam prior to reassembly. Clean and grease the pivot stub which is set in the backplate.

10 Check the condition of the felt seal which prevents the escape of grease from the end of the camshaft. If it is in any way damaged or too dirty to be of further use, it must be renewed before the shaft is relocated in the backplate. Dip the seal in clean engine oil before refitting it, pack grease into the backplate passage and refit the camshaft. Fit the wear indicator over the camshaft splines, aligning its inner tab with the camshaft punch mark. Rotate the camshaft to its correct position and refit the operating arm, aligning its punch mark with that of the camshaft. Tighten the pinch bolt to a torque setting of 0.8 - 1.2 kgf m (6 - 9 lbf ft).

11 Before refitting existing shoes, roughen the lining surface sufficiently to break the glaze which will have formed in use. Glasspaper or emery cloth is ideal for this purpose but take care not to inhale any of the asbestos dust that may come from the lining surface.

12 Fitting the brake shoes and springs to the brake backplate is a reversal of the removal procedure. Some patience will be needed to align the assembly with the pivot and operating cam whilst still retaining the springs in position; once the brake shoes are correctly aligned, they can be pushed back into position by pressing downwards to snap them into position. Do not use excessive force, as there is risk of distorting them permanently.

11.4 Examine carefully brake drum surface – removal all traces of dirt and grease

11.7 Removing brake shoes from backplate

Tyre changing sequence - tubed tyres

 Deflate tyre. After pushing tyre beads away from rim flanges push tyre bead into well of rim at point opposite valve. Insert tyre lever adjacent to valve and work bead over edge of rim.

Use two levers to work bead over edge of rim. Note use of rim protectors

 Remove inner tube from tyre

When first bead is clear, remove tyre as shown

 When fitting, partially inflate inner tube and insert in tyre

Work first bead over rim and feed valve through hole in rim. Partially screw on retaining nut to hold valve in place.

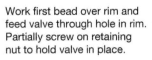

 Check that inner tube is positioned correctly and work second bead over rim using tyre levers. Start at a point opposite valve.

Work final area of bead over rim whilst pushing valve inwards to ensure that inner tube is not trapped

12 Tyres: removal, repair and refitting

1   To remove the tyre from either wheel, first detach the wheel from the machine. Deflate the tyre by removing the valve core, and when the tyre is fully deflated, push the bead away from the wheel rim on both sides so that the bead enters the centre well of the rim. Remove the locking ring and push the tyre valve into the tyre itself, then remove its nut and washer and press in also the security bolt (where fitted).

2   Insert a tyre lever close to the valve and lever the edge of the tyre over the outside of the rim. Very little force should be necessary; if resistance is encountered it is probably due to the fact that the tyre beads have not entered the well of the rim, all the way round. Damage to the soft alloy rim by tyre levers can be prevented by the use of plastic rim protectors.

3   Once the tyre has been edged over the wheel rim, it is easy to work round the wheel rim, so that the tyre is completely free from one side. At this stage the inner tube and security bolt can be removed.

4   Now working from the other side of the wheel, ease the other edge of the tyre over the outside of the wheel rim that is furthest away. Continue to work around the rim until the tyre is completely free from the rim.

5   If a puncture has necessitated the removal of the tyre, reinflate the inner tube and immerse it in a bowl of water to trace the source of the leak. Mark the position of the leak, and deflate the tube. Dry the tube, and clean the area around the puncture with a petrol soaked rag. When the surface has dried, apply rubber solution and allow this to dry before removing the backing from the patch, and applying the patch to the surface.

6   It is best to use a patch of self vulcanizing type, which will form a permanent repair. Note that it may be necessary to remove a protective covering from the top surface of the patch after it has sealed into position. Inner tubes made from a special synthetic rubber may require a special type of patch and adhesive, if a satisfactory bond is to be achieved.

7   Before replacing the tyre, check the inside to make sure that the article that caused the puncture is not still trapped inside the tyre. Check the outside of the tyre, particularly the tread area to make sure nothing is trapped that may cause a further puncture.

8   If the inner tube has been patched on a number of past occasions, or if there is a tear or large hole, it is preferable to discard it and fit a replacement. Sudden deflation may cause an accident.

9   To replace the tyre, inflate the inner tube for it just to assume a circular shape but only to that amount, and then push the tube into the tyre so that it is enclosed completely. Lay the tyre on the wheel at an angle, and insert the valve through the rim tape and the hole in the wheel rim. Attach the locking ring on the first few threads, sufficient to hold the valve captive in its correct location.

10 Starting at the point furthest from the valve, push the tyre bead over the edge of the wheel rim until it is located in the central well. Continue to work around the tyre in this fashion until the whole of one side of the tyre is on the rim. It may be necessary to use a tyre lever during the final stages. Insert the security bolt with its washer and nut.

11 Make sure there is no pull on the tyre valve and again commencing with the area furthest from the valve, ease the other bead of the tyre over the edge of the rim. Finish with the area close to the valve, pushing the valve up into the tyre until the locking ring touches the rim. This will ensure that the inner tube is not trapped when the last section of bead is edged over the rim with a tyre lever.

12 Check that the inner tube is not trapped at any point. Reinflate the inner tube, check that the tyre is seating correctly around the wheel rim. There should be a thin rib moulded around the wall of the tyre on both sides, which should be an equal distance from the wheel rim at all points. If the tyre is unevenly located on the rim, try bouncing the wheel when the tyre is at the recommended pressure. It is probable that one of the beads has not pulled clear of the centre well. When the tyre is correctly fitted and fully inflated tighten down hard the security bolt retaining nut.

13 Always run the tyres at the recommended pressures and never under or over inflate. The correct pressures are given in the Specifications Section of this Chapter.

14 Tyre replacement is aided by dusting the side walls, particularly in the vicinity of the beads, with a liberal coating of french chalk. Washing up liquid can also be used to good effect.

15 Never replace the inner tube and tyre without the rim tape in position, (MTX125/200 models only). If this precaution is overlooked there is a good chance of the ends of the spoke nipples chafing the inner tube and causing a crop of punctures.

16 Never fit a tyre that has a damaged thread or sidewalls. Apart from legal aspects, there is a very great risk of a blowout, which can have very serious consequences on a two wheeled vehicle.

13 Valve cores and caps: general

1   Valve cores seldom give trouble, but do not last indefinitely. Dirt under the seating will cause a puzzling 'slow-puncture'. Check that they are not leaking by applying spittle to the end of the valve and watching for air bubbles.

2   A valve cap is a safety device, and should always be fitted. Apart from keeping dirt out of the valve, it provides a second seal in case of valve failure, and may prevent an accident resulting from sudden deflation.

14 Wheel balancing

1   It is customary to balance the wheels complete with tyres and tube. The out of balance forces which exist are eliminated and the handling of the machine is improved in consequence. A wheel which is badly out of balance produces through the steering a most unpleasant hammering effect at high speeds.

2   Some tyres have a balance mark on the sidewall, usually in the form of a coloured spot. This mark must be in line with tyre valve, when the tyre is fitted to the inner tube. Even then the wheel may require the addition of balance weights, to offset the weight of the tyre valve itself.

3   If the wheel is raised clear of the ground and is spun, it will probably come to rest with the tyre valve or the heaviest part downward and will always come to rest in the same position. Balance weights must be added to a point diametrically opposite this heavy spot until the wheel will come to rest in ANY position after it is spun.

4   Although balance weights are not made available specifically for these models, most good motorcycle dealers or tyre fitting specialists will be able to provide weights in a range of sizes. On machines equipped with Comstar wheels, the weights are clipped to the rim central flange, while on wire-spoked wheels, the weights are clamped around the head of a convenient spoke.

Chapter 7 Electrical system

Contents

Specifications

Unless otherwise stated, information applies to all models

Electrical system

Voltage	12
Earth	Negative
Alternator output – @ 5000 rpm:	
MBX125	130W
MTX125	90W
MTX200	120W

Battery

Make	Yuasa
Type:	
MBX125	YB5L-B
MTX125/200	YB3L-A
Capacity:	
MBX125	5Ah
MTX125/200	3Ah

Fuse

Rating:	
MBX125	15A
MTX125/200	10A

Charging system

Charging begins at	1600 rpm	
Charging output:	**MBX125**	**MTX125/200**
@ 3000 rpm	0.5A minimum/18.0V	2.2A minimum/18.0V
@ 8000 rpm	15.0A maximum/18.2V	4.7A maximum/18.2V
No load regulated voltage – MBX125	14-15V	
AC regulator minimum resistance – MTX125/200	100 K ohm	

Alternator

Source coil resistance – MBX125	0.3 – 0.7 ohm (Yellow – Yellow wires)
Charging coil resistance – MTX125/200	0.79 – 1.10 ohm (Yellow – Pink wires)
Lighting coil resistance – MTX125/200	0.54 – 0.72 ohm (White/Yellow wire – earth)

Bulbs

Headlamp	12V, 60/55W
Pilot or parking lamp	12V, 4W
Stop/tail lamp	12V, 21/5W
Turn signal lamps	12V, 21W
Main beam warning lamp	12V, 1.7W
All other warning lamps, instrument illuminating lamps	12V, 3.4W

Torque wrench settings

Component	kgf m	lbf ft
Battery case mounting bolts	0.9 – 1.3	6.5 – 9.5
Temperature gauge sender unit	0.8 – 1.4	6 – 10
Alternator rotor retaining nut	6.0 – 7.0	43 – 50.5
Headlamp nacelle mounting bolts – MTX125/200	0.8 – 1.2	6 – 9

1 General description

On MBX125 models the power for the electrical system is provided by a three-phase alternator mounted on the crankshaft left-hand end, the output from which is rectified and controlled by an electronic regulator/rectifier unit before being passed to the battery and main electrical circuit.

MTX125/200 models employ a more basic system in which an AC generator (alternator) is mounted on the crankshaft left-hand end. Part of its output, from the charging coils, is rectified and controlled by a simpler form of electronic regulator/rectifier unit before being passed

to the battery to provide power for the ancillary electrical components. When the main lighting switch is in the 'H' or 'P' positions current is fed to the pilot lamp, tail lamp and instrument illuminating lamps from the battery. The remainder of the generator output, from the lighting coil, is fed direct to the headlamp and main beam warning lamp bulbs via the dip switch and main lighting switch, so that the headlamp will only light when the main lighting switch is in the 'H' position and the engine is running. An AC regulator provides a simple form of voltage control by soaking up and dispersing as heat, any surplus current generated, for example when the headlamp is not switched on but the engine is running.

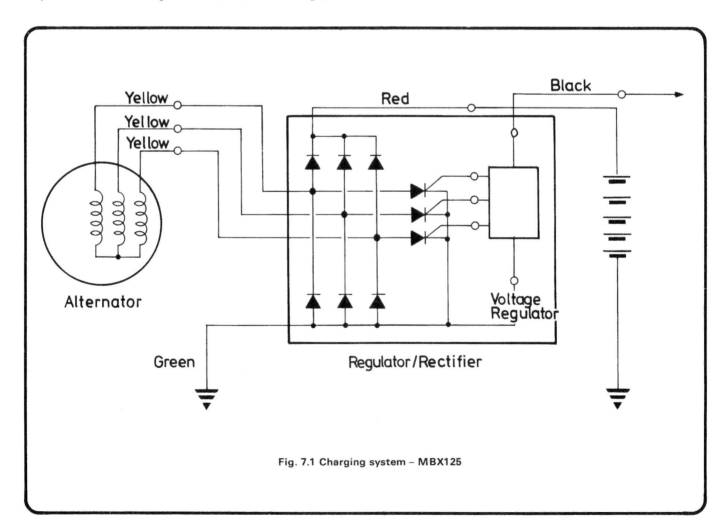

Fig. 7.1 Charging system – MBX125

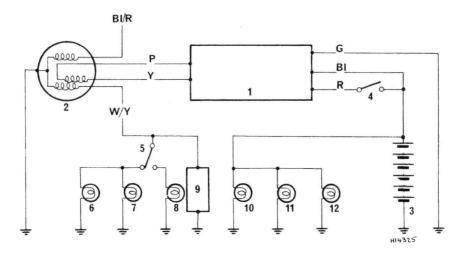

Fig. 7.2 Charging system – MTX125 and 200

P	Pink	Bl/R	Black and red	3	Battery	8	Headlamp dip beam
Y	Yellow	W/Y	White and yellow	4	Ignition switch	9	AC Regulator
G	Green			5	Dip switch	10	Tail lamp
R	Red	1	Regulator/rectifier	6	Main beam warning light	11	Instrument light
Bl	Black	2	Generator	7	Headlamp main beam	12	Pilot lamp

2 Electrical system: general information and preliminary checks

1   In the event of an electrical system fault, always check the physical condition of the wiring and connectors before attempting any of the test procedures described here and in subsequent Sections. Look for chafed, trapped or broken electrical leads and repair or renew these as necessary. Leads which have broken internally are not easily spotted, but may be checked using a multimeter or a simple battery and bulb circuit as a continuity tester. This arrangement is shown in the accompanying illustratiom. The various multi-pin connectors are generally trouble-free but may corrode if exposed to water. Clean them carefully, scraping off any surface deposits, and pack with silicone grease during assembly to avoid recurrent problems. The same technique can be applied to the handlebar switches.

2   A sound, fully charged battery is essential to the normal operation of the system. There is no point in attempting to locate a fault if the battery is partly discharged or worn out. Check battery condition and recharge or renew the battery before proceeding further.

3   Many of the test procedures described in this Chapter require that voltages or resistances be checked. This necessitates the use of some form of test equipment such as a simple and inexpensive multimeter of the type sold by electronics or motor accessory shops.

4   If you doubt your ability to check the electrical system entrust the work to a Honda Service Agent. In any event have your findings double checked before consigning expensive components to the scrap bin.

2.1 Electrical faults are often due to corrosion in connectors – check carefully

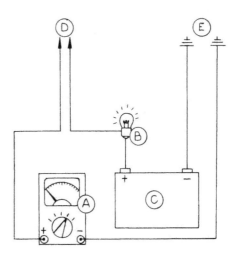

Fig. 7.3 Simple testing arrangement for checking the wiring

A	Multimeter	D	Positive probe
B	Bulb	E	Negative probe
C	Battery		

3 Battery: examination and maintenance

1   Details of the regular checks needed to maintain the battery in good condition are given in Routine Maintenance, together with instructions on removal and refitting and general battery care. Batteries can be dangerous if mishandled; read carefully the 'Safety First' section at the front of this Manual before starting work, and always wear overalls or old clothing in case of accidental acid spillage. If acid is ever allowed to splash into your eyes or on to your skin, flush it away with copious quantities of fresh water and seek medical advice immediately.
2   When new, the battery is filled with an electrolyte of dilute sulphuric acid having a specific gravity of 1.280 at 20°C (68°F). Subsequent evaporation, which occurs in normal use, can be compensated for by topping up with distilled or demineralised water only. Never use tap water as a substitute and do not add fresh electrolyte unless spillage has occurred.
3   The state of charge of a battery can be checked using an hydrometer.
4   The normal charge rate for a battery is $1/10$ of its rated capacity, thus for a 5 ampere hour unit charging should take place at 0.5 amp. Exceeding this figure could cause the battery to overheat, buckling the plates and rendering it useless. Few owners will have access to an expensive current controlled charger, so if a normal domestic charger is used check that after a possible initial peak, the charge rate falls to a safe level. If the battery becomes hot during charging **stop**. Further charging will cause damage. Note that cell caps should be loosened and vents unobstructed during charging to avoid a build-up of pressure and risk of explosion.
5   After charging, top up with distilled water as required, then check that specific gravity and battery voltage. Specific gravity should be above 1.270 and a sound, fully charged battery should produce 15 – 16 volts. If the recharged battery discharges rapidly if left disconnected it is likely that an internal short caused by physical damage or sulphation has occurred. A new battery will be required. A sound item will tend to lose its charge at about 1% per day.

4 Charging system: checking the output

MBX125

1   With the machine on its centre stand, start the engine and warm it up for about 10 minutes. Stop the engine and remove the seat, the side panels and the fuel tank to reach the regulator/rectifier unit block connector. Disconnect the black wire at the connector but leave the other wires connected; this is best achieved by carefully prising the terminal out of one half of the connector, which can then be plugged together again. Disconnect the battery positive ($+$) wire from its terminal. Note the battery must be fully charged and in good condition if the test is to be accurate.
2   Connect a suitable voltmeter across the battery terminals (meter negative ($-$) lead to battery negative ($-$) terminal) and an ammeter between the battery positive ($+$) terminal and the end of the disconnected wire, then check that the main lighting switch is in the off position.
3   Start the engine and increase speed, comparing the readings obtained at the two specified engine speeds with those given in the Specifications Section of this Chapter. As soon as the various readings have been noted, stop the engine; the test should be performed as quickly as possible to prevent the risk of damage to the engine or to the electrical components. Do not forget to reconnect the black wire at the regulator/rectifier block connector.
4   If the readings obtained differ significantly from those specified, and if the battery is known to be fully charged, the charging system is faulty. The alternator (source coil), the regulator/rectifier unit, the wiring and switches must be checked as described in the relevant Sections of this Chapter to eliminate the fault.
5   To check the performance of the regulator part of the regulator/rectifier unit, no preliminary dismantling is necessary except to remove the right-hand side panel to gain access to the battery terminals; again, note that the battery must be fully charged for the test to be accurate. Connect a voltmeter across the battery terminals (meter negative ($-$) lead to battery negative ($-$) terminal) and start the engine, having checked that the lights are switched off.
6   Increase engine speed until an indicated output of 14 volts is reached then increase speed further and check that the reading does

not exceed 15 volts; again, perform this test as quickly as possible to prevent the risk of damage. If the reading obtained is significantly above (or below) 14–15 volts the regulator is faulty and should be renewed, although if it appears to be in good condition as a result of the resistance test given in Section 6 of this Chapter, the switches, wiring, connectors and regulator mountings (earth connections) should be checked first.

MTX125/200

7   Proceed exactly as described in paragraphs 1–4 above to test the output of the charging system, referring to the readings given in the Specifications Section of this Chapter to check those obtained.
8   The condition of the AC regulator unit can be checked only by the the test described in Section 6 of this Chapter.

5 Alternator: testing

1   Since no other readings are provided, the alternator can be tested only by checking the condition of its source coils, as described below.
2   Start by disconnecting the alternator wires from the main loom at the connectors near the top of the frame front downtube. It may be necessary to remove the radiator shroud(s) and to slacken the radiator mountings to permit this. An ohmmeter, or a multimeter set to the appropriate resistance scale, will be necessary to make the tests.
3   On MBX125 models, measure the resistance between all three yellow wires, comparing the readings obtained with that given in the Specifications Section of this Chapter, then measure the resistance between each yellow wire terminal and a good earth point on the crankcase; a reading of almost infinite resistance should be obtained, indicating that each coil is completely insulated.
4   If one or more of the yellow wires is isolated from its fellows, or if one of the wires is shorting to earth, the stator is faulty and must be renewed. See below.
5   On MTX125/200 models make the meter connections as shown in the Specifications Section of this Chapter, comparing the readings obtained with those given, then measure the resistance between the yellow and pink wire terminals and a good earth point on the crankcase to check that the charging coil is completely insulated; a reading of almost infinite resistance should be obtained.
6   If any test produces results different from those given, the coil is defective and must be renewed, but note that where a specified resistance should be measured, if a reading of infinite resistance (indicating an open circuit or broken wire) or of very little resistance (indicating a short circuit or trapped wire) is obtained, the fault may be repaired by the owner.
7   On all models the coils are part of the stator plate assembly and cannot be obtained individually. It may be possible for an auto-electrical specialist to rewind a faulty coil, saving much of the cost of a new assembly. The only other alternative is to find a good used component at a motorcycle breakers.
8   The alternator components are removed and refitted as described in Chapter 1.

6 Regulator/rectifier unit: location and testing

Regulator/rectifier unit: location and testing

1   The regulator/rectifier is a sealed, heavily-finned unit mounted on the left-hand side of the frame lower top tube, immediately to the rear of the steering head. Remove the seat, both side panels and the fuel tank to gain access to it.
2   To test the unit, disconnect it at the multi-pin connector joining it to the main wiring loom; the test must be carried out using a Sanwa SP-10D electrical tester set to the kilo ohms scale, making the connections as shown in the accompanying chart. If the correct equipment is not available some idea of the unit's condition may be gained using another multimeter or similar, the resistance scale being determined by trial and error. If the unit is tested with anything other than the specified tester, the results should be confirmed by a Honda Service Agent before any firm conclusions are reached.
3   If the results obtained differ from those specified at any point, the unit is faulty and must be renewed; repairs are not possible. Note the performance test of the regulator part of the unit given in Section 4 of this Chapter.

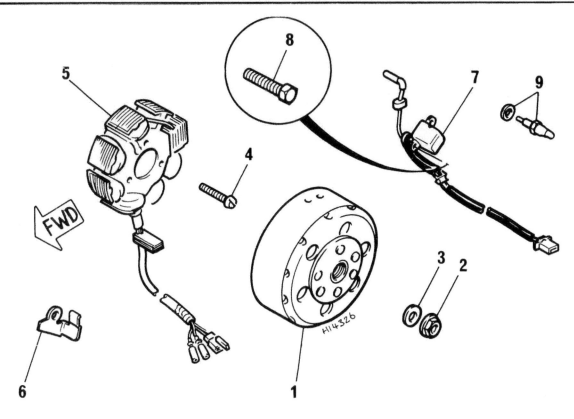

Fig. 7.4 Alternator assembly

1 Rotor	4 Screw – 3 off	6 Clamp	8 Bolt – 2 off
2 Nut	5 Stator	7 Pulser coil	9 Neutral switch
3 Washer			

Regulator/rectifier unit – MTX125/200

1 The charging system regulator/rectifier unit is a sealed metal unit mounted on the right-hand side of the frame top tubes, immediately to the rear of the steering head. Remove the seat, both side panels, the radiator shrouds and the fuel tank to gain access to it.

5 To test the unit, disconnect it at the two multi-pin block connectors joining it to the main wiring loom; the test must be carried out using either a Sanwa SP-10D electrical tester set to the kilo ohms scale, or a Kowa TH-5H electrical tester set to the X100 ohm scale. Measure the resistance between the pairs of wires as shown in the accompanying test table, comparing the readings obtained with those specified. If the specified equipment is not available some idea of the unit's condition may be gained by using an ordinary multimeter or ohmmeter, determining by trial and error the correct resistance range; the results of any such test should be confirmed by a Honda Service Agent using the correct equipment before any firm conclusions are reached.

6 If the results obtained differ from those specified at any point, the unit is faulty and must be renewed; repairs are not possible.

AC regulator – MTX125/200

7 The regulator controlling the voltage in the lighting circuit is a sealed heavily-finned metal unit mounted on the left-hand side of the frame top tubes, immediately to the rear of the steering head. Remove the seat, both side panels, the radiator shrouds and the fuel tank to gain access to it. Note that the unit must be kept as clean as possible at all times and that regular checks should be made to ensure that its mounting bolt and connections are free from corrosion and securely fastened.

8 If the unit is faulty, the excessive voltage in the system will cause the bulbs to blow, their filaments having melted ends. Before suspecting the AC regulator check the main lighting switch, the dip switch and the headlamp and main beam warning lamp bulb connections.

9 To test the unit itself disconnect the two wires and measure the resistance between them with an ohmmeter or a multimeter set to the appropriate resistance scale; a reading of anywhere between 100 kilo ohms to infinite resistance should be obtained. If a reading of less than 100 kilo ohms is obtained, the unit is faulty and must be renewed; repairs are not possible.

6.1 Regulator/rectifier unit – MBX125

6.4 Regulator/rectifier unit – MTX125/200

6.7 AC regulator – MTX125/200

Unit kΩ

Tester (–) \ Tester (+)	Yellow	Yellow	Yellow	Black	Red	Green
Yellow		∞	∞	∞	0.5 –10	∞
Yellow	∞		∞	∞	0.5 –10	∞
Yellow	∞	∞		∞	0.5 –10	∞
Black	30-70	30-70	30-70		30 –100	30-70
Red	∞	∞	∞	∞		∞
Green	0.5-10	0.5-10	0.5-10	1-20	2 –50	

HI4327

Fig. 7.5 Regulator/rectifier unit test table – MBX125

Tester (–) \ Tester (+)	Yellow	Pink	Green	Red	Black
Yellow		∞	∞	1-20	∞
Pink	∞		∞	1-20	∞
Green	1-20	1-20		3-100	0.2-20
Red	∞	∞	∞		∞
Black	1-50	1-50	0.2-10	3-100	

HI4328

Fig. 7.6 Regulator/rectifier unit test table – MTX125 and 200

7 Switches: general

1   While the switches should give little trouble, they can be tested using a multimeter set to the resistance function or a battery and bulb test circuit. Using the information given in the wiring diagram at the end of this Manual, check that full continuity exists in all switch positions and between the relevant pairs of wires. When checking a particular circuit follow a logical sequence to eliminate the switch concerned.

2   As a simple precaution always disconnect the battery before removing any of the switches, to prevent the possibility of a short circuit. Most troubles are caused by dirty contacts, but in the event of the breakage of some internal part, it will be necessary to renew the complete switch.

3   If a switch is tested and found to be faulty, there is nothing to be lost by attempting a repair. It may be that worn contacts can be built up with solder, or that a broken wire terminal can be repaired, again using a soldering iron. The handlebar switches may be dismantled to a certain extent. It is, however, up to the owner to decide if he has the skill to carry out this sort of work.

4   While none of the switches require routine maintenance, some regular attention will prolong their life. The regular and constant application of WD40 or a similar water-dispersant spray not only prevents problems occurring due to water-logged switches and the resulting corrosion, but also makes the switches much easier and more positive to use. Alternatively, the switch may be packed with a silicone-based grease to achieve the same result.

8 Fuse: location and renewal

1   The electrical system is protected by a single fuse which is retained in a plastic casing set in the battery positive (+) terminal lead, and is clipped to a holder next to the battery. If the spare fuse is ever used, replace it with one of the correct rating as soon as possible.

2   Before renewing a fuse that has blown, check that no obvious short circuit has occurred, otherwise the replacement fuse will blow immediately it is inserted. It is always wise to check the electrical circuit thoroughly, to trace the fault and eliminate it.

3   When a fuse blows while the machine is running and no spare is available, a 'get you home' remedy is to remove the blown fuse and wrap it in silver paper before replacing it in the fuse holder. The silver paper will restore the electrical continuity by bridging the broken fuse wire. This expedient should never be used if there is evidence of short circuit or other major electrical faults, otherwise more serious damage will be caused. Replace the 'doctored' fuse at the earliest possible opportunity, to restore full circuit protection.

9 Bulbs: renewal

1   To renew the headlamp or parking/pilot lamp bulb, remove the fairing (MBX125) or headlamp nacelle (MTX125/200) and remove the two bolts which retain the headlamp assembly. Partially withdraw the assembly from its bracket and unplug the headlamp bulb connector. On MBX models remove the rubber cover, then disengage the spring clip and withdraw the bulb. On MTX models, peel back the rubber cover and turn the bulbholder anticlockwise to gain access to the bulb.

2   The headlamp bulb is a quartz-halogen type with a conventional H4 fitting; do not touch the bulb's glass envelope as skin acids will shorten the bulb's service life. If the bulb is touched accidentally, it should be wiped carefully when cold with a rag soaked in methylated spirits and dried before being refitted. On refitting, note that the locating tangs on the bulb metal collar are offset so that the bulb can only be refitted in the correct position.

3   The parking/pilot lamp bulb assembly can be pulled out of its rubber grommet in the reflector; the bulb should be pressed in and twisted anti-clockwise to release it.

4   All instrument panel bulbs are of the capless type, being pressed into bulb holders which are rubber-mounted in the base of the instrument or the instrument panel. Refer to Chapter 5 for instructions on the work necessary to gain access to the bulbs; be careful not to damage the delicate wire terminals on removing and refitting the bulbs.

5   All turn signal lamp bulbs and the stop/tail lamp bulb are of the bayonet type and can be released by pushing in, turning anti-clockwise and pulling from the holder. The stop/tail lamp is fitted with a twin-filament bulb which has offset pins to prevent accidental reversal in its holder.

6   Except for the front turn signal lamps on MBX125 machines, all lamp lenses are retained by one or two screws; take care not to tear the lens seal on removal. On refitting, clean away any moisture or corrosion from the bulb and holder, check that the holder contacts are free to move against spring pressure and that the lens seal is correctly fitted and remains in place as the lens is refitted. Do not overtighten the lens retaining screw(s), the lens will crack if over-stressed.

7   On MBX125 models, to renew the front turn signal lamp bulb(s), first remove the fairing then unscrew the retaining bolt and screw which secure the lamp assembly mounting bracket to the inside of the fairing. Withdraw the lamp and peel away the soft rubber gasket to separate the lens from the reflector. The bulb can then be withdrawn and the lamp assembly checked as described in paragraphs 5 and 6 above. On refitting, fit the bulb into the holder and place the lens on the reflector. Refit the rubber gasket, taking care not to stretch or damage it and place the assembly on the fairing inside surface. Hook the mounting bracket front end into the fairing slot and tighten carefully the retaining screw and bolt.

9.1a Remove two bolts to release headlamp assembly ...

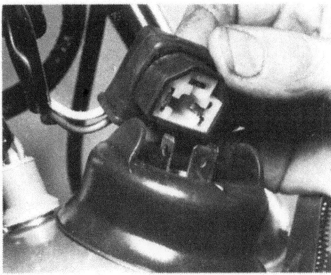

9.1b ... unplug headlamp bulb connector ...

9.1c ... and remove rubber cover to expose bulb mounting

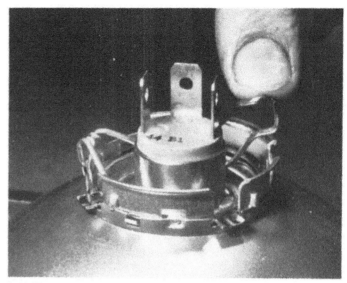

9.1d Disengage spring clip to release bulb

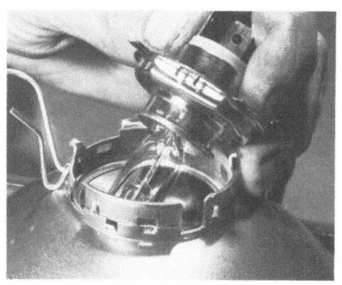

9.2a Tangs on bulb collar will fit only one way in headlamp assembly

9.2b Headlamp beam horizontal alignment is adjusted by rotating spring-loaded screw

9.3 Parking/pilot lamp bulb holder is a press fit in headlamp assembly

9.4 Instrument panel bulbs are pressed into base of instruments or instrument panel

9.5a Lamp lenses are retained by single screw and clip ...

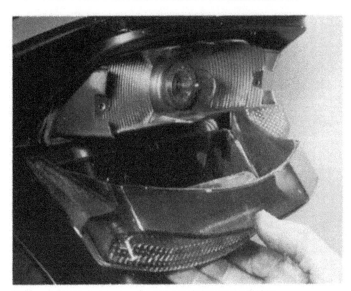

9.5b ... or by two screws – do not overtighten on refitting

9.7a Front turn signal lamp bulb renewal – MBX125 – remove bracket retaining screw and bolt ...

9.7b ... and disengage bracket from front mounting slot

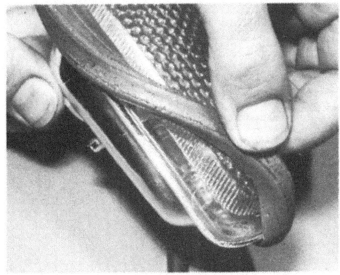

9.7c Peel off foam rubber gasket ...

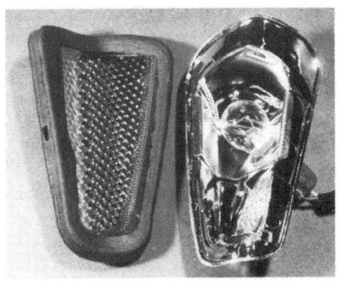

9.7d ... to separate lens from lamp reflector – bulb can then be removed

10 Horn: locating and testing

1   The horn is mounted on the bottom yoke or on the headlamp bracket via a flexible metal strip retained by a single bolt. No maintenance is required other than regular cleaning to remove road dirt and occasional spraying with WD40 or a similar water dispersant spray to minimise internal corrosion.
2   If the horn fails to work, first check that the battery is fully charged. If full power is available, a simple test will reveal whether the current is reaching the horn. Disconnect the horn wires and substitute a 12 volt bulb. Switch on the ignition and press the horn button. If the bulb fails to light, check the horn button and wiring as described elsewhere in this Chapter. If the bulb does light, the horn circuit is proved good and the horn itself must be checked.
3   With the horn wires still disconnected, connect a fully charged 12 volt battery directly to the horn. If it does not sound, a sharp tap on the outside may serve to free the internal contacts. If this fails, the horn must be renewed as repairs are not possible.
4   Different types of horn may be fitted; if a screw and locknut is provided on the outside of the horn, the internal contacts may be adjusted to compensate for wear and to cure a weak or intermittent horn note. Slacken the locknut and rotate slowly the screw until the clearest and loudest note is obtained, then retighten the locknut. If no means of adjustment is provided on the horn fitted, it must be renewed.

10.1 Horn is mounted on fork bottom yoke or on headlamp bracket

11 Turn signal relay: location and testing

1   The turn signal relay is a sealed black plastic box, rubber-mounted to protect it from vibration. It is mounted on the frame top tubes, in front of the fuel tank rear mounting on MBX125 machines and immediately behind the steering head on MTX125/200 models. It will be necessary to remove the seat, both side panels and the fuel tank to gain access to it.
2   If the turn signal lamps cease to function correctly, there may be any one of several possible faults responsible which should be checked before the relay is suspected. First check that the lamps are correctly mounted and that all the earth connections are clean and tight. Check that the bulbs are of the correct wattage and that corrosion has not developed on the bulbs or in their holders. Any such corrosion must be thoroughly cleaned off to ensure proper bulb

contact. Also check that the turn signal switch is functioning correctly and that the wiring is in good order. Finally ensure that the battery is fully charged.
3   Faults in any one or more of the above items will produce symptoms for which the turn signal relay may be blamed unfairly. If the fault persists even after the preliminary checks have been made, the relay must be at fault. Unfortunately the only practical method of testing the relay is to substitute a known good one. If the fault is then cured, the relay is proven faulty and must be renewed.

12 Low oil level warning lamp circuit: testing

1   The circuit consists of a float-type switch mounted in the top of the oil tank and the bulb itself, which is mounted in the instrument panel (MBX125) or in the speedometer (MTX125/200).
2   When the ignition is switched on the lamp should light for approximately 5 seconds as a check that the system is working, then it will go out. If it does not go out there is insufficient (200 cc/0.35 pint or less) oil in the tank; top up the tank as described in Routine Maintenance and check again that the lamp goes out. If the lamp does not light at all the system is faulty and must be checked.
3   The most likely cause of failure will be a blown bulb, but it is worth checking first that the battery is fully charged; check that all other systems are working normally as a quick test of this. The bulb is removed and refitted as described in Section 9 of this Chapter and Section 13 of Chapter 5.
4   If the bulb is in good condition the fault must lie in the switch or in the wiring. Remove the seat and side panels to gain access to the oil tank top surface and disconnect the switch at the multi-pin block connector joining it to the main loom. Test the wiring in the main loom as follows.
5   First check that full battery voltage is available, connecting a DC voltmeter between the black wire terminal and a suitable earth point on the frame and switching on the ignition. If battery voltage is not measured check the black wire back to the ignition switch and the switch itself. Next use a multimeter set to one of the resistance scales to check that there is continuity between the green wire terminal and a good earth point on the frame, and between the green/red wire terminal and the warning lamp bulb holder; if resistance is measured in either case the wire is damaged or broken and must be checked carefully until the fault is found.
6   If the bulb and wiring are in good condition the fault must lie in the switch. Unplug it from the oil tank and wash off all traces of oil, checking that the terminals are clean and making good contact. Use a multimeter set to the appropriate resistance scale to test the switch. With the float at the bottom of the switch column, measure the resistance between the switch green/red and black wire terminals; a reading of 5 – 15 ohms should be obtained. Check that there is no continuity (ie infinite resistance) between the green and black wire terminals when the float is fully lowered. Finally lift the float fully to the top of the column and measure again the resistance between the green/red and black wire terminals; the reading should now be 340 ohms.
7   If any of the above tests reveals a fault in the switch, or if obvious signs of damage can be seen on removing the switch from the oil tank, the switch must be renewed; repairs are not possible.
8   To test the accuracy of the switch, remove it from the tank and clean it, as described above, then connect it again to the main loom, switch on the ignition, raise the float to the top of the column and wait for the lamp to go out. When the lamp goes out move the float to the bottom of the column, whereupon the lamp should light again, and slowly move the float back up the column until the lamp goes out. Measure the distance between the float stop at the bottom of the column and the bottom of the float; the distance should be 4 – 5 mm (0.18 in) although a tolerance of 1.0 mm (0.04 in) above or below this figure is allowed. If the lamp lights above or below these limits the switch is inaccurate.
9   The only solution to an inaccurate switch is to renew it, although it must be up to the owner to decide whether this is justified; it is a simple matter to drain the oil tank and to pour in oil until the lamp goes out, measuring carefully and recording for future reference the amount of oil necessary to do this.

11.1a Location of turn signal relay – MBX125 ...

11.1b ... and MTX125/200

12.1 Oil level warning switch is mounted in top of oil tank

given in paragraph 5 below. If the needle's movement is still faulty, or if it does not move at all, the gauge or wiring is faulty and must be checked until the fault is eliminated. Proceed as described below from paragraph 8 onwards.

5   To test fully the sender unit remove it from the machine; it can be checked by measuring its resistance at various temperatures. The task must be carried out in a well-ventilated area and great care must be taken to avoid the risk of personal injury. The equipment necessary is a heatproof container, a small gas-powered camping stove or similar, a thermometer capable of reading up to 120°C (250°F) and an ohmrneter or a multimeter set to the appropriate resistance scale.

6   Fill the container with oil and suspend the sender unit on some wire so that the probe end is immersed in it. Connect one of the meter leads to the unit body and the other to its terminal. Suspend the thermometer so that the bulb is close to the sender probe.

7   Start to heat the oil, and make a note of the resistance reading at each temperature shown in the table below. If the unit does not give readings which approximate quite closely to those shown, it must be renewed.

8   To test the gauge assembly, dismantle the instrument panel as described in Section 13 of Chapter 5 until the gauge terminals are exposed. Remove the three nuts and washers and disconnect the wires from their terminals, then test the wiring as follows.

9   Switch on the ignition and use a DC voltmeter to check that full battery voltage is available, connecting the meter between the black wire terminal and a suitable earth point on the frame. If battery voltage is not available, the fault must be in the black wire or in the ignition switch; check these carefully until the fault is located. Next use an ohmmeter or a multimeter set to the resistance scale to check that there is continuity between the green wire terminal and a good earth point on the frame; if this is not the case check the wire until the break is located and repaired. Finally check the green/blue wire; first check that it is not broken by connecting the meter probe to the terminals at each end (gauge assembly and sender unit) of the wire and checking for continuity, then with the wire disconnected at both ends, connect the meter probes between one of the wire terminals and a good earth point on the frame. No continuity (ie infinite resistance) should be measured; if continuity exists, the wire is trapped or pinched somewhere along its length thus creating a short circuit.

10 If the wiring and sender unit are tested and found to be in good condition, the fault must be in the gauge assembly, which can only be renewed as repairs are not possible. On MBX125 machines the gauge can be renewed individually, but on MTX125/200 models it forms a part of the sealed speedometer assembly; the complete unit must be renewed to cure a gauge fault.

13 Water temperature gauge circuit: testing

1   The circuit consists of the sender unit mounted in the cylinder head water jacket and the gauge assembly mounted in the instrument panel (MBX125) or in the speedometer (MTX125/200).

2   The sender unit is removed and refitted as described in Section 9 of Chapter 2, while the gauge assembly is removed and refitted as described in Section 13 of Chapter 5.

3   To test the system, first ensure that the battery is fully charged by checking that all other electrical circuits work properly, then disconnect its wire from the sender unit and check that the ignition is switched on. The temperature gauge needle should point to 'C'. Earth the sender unit wire on the cylinder head, whereupon the needle should immediately swing over to 'H'. Do not earth the wire for more than 5 seconds or the gauge may be damaged.

4   If the needle moves as described, the sender unit is proven faulty and must be renewed, although a fuller check of its performance is

13.1a Water temperature gauge circuit consists of sender unit mounted in cylinder head ...

13.1b ... and gauge unit mounted in instrument panel – MBX125 shown

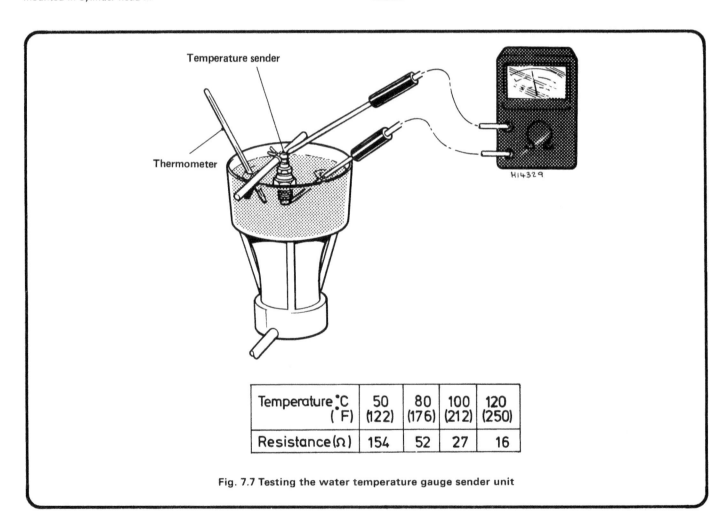

Temperature °C (°F)	50 (122)	80 (176)	100 (212)	120 (250)
Resistance (Ω)	154	52	27	16

Fig. 7.7 Testing the water temperature gauge sender unit

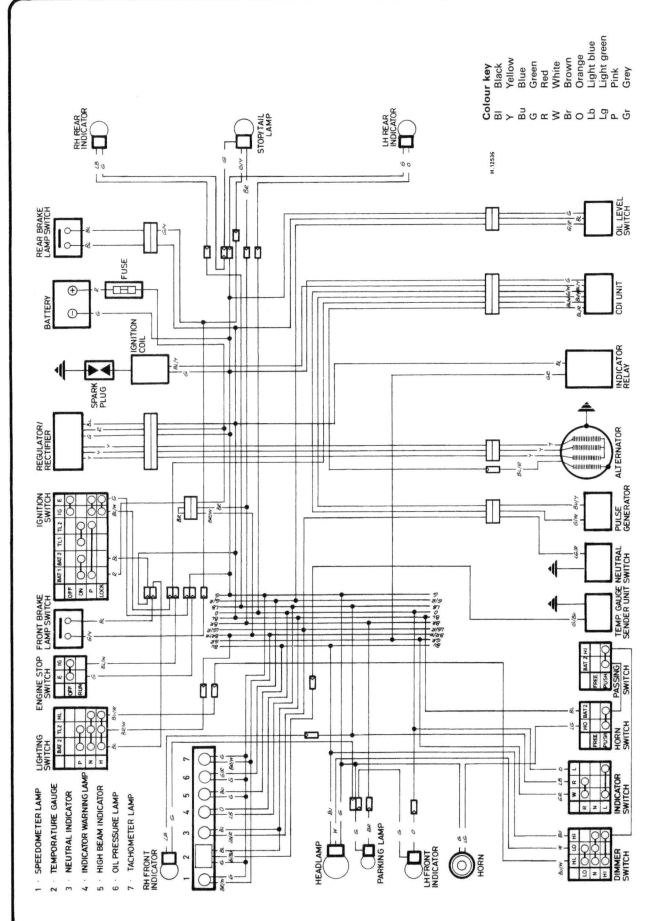

Colour key

Bl	Black
Y	Yellow
Bu	Blue
G	Green
R	Red
W	White
Br	Brown
O	Orange
Lb	Light blue
Lg	Light green
P	Pink
Gr	Grey

H.12536

Wiring diagram – MBX125 model

1 · SPEEDOMETER LAMP
2 · TEMPERATURE GAUGE
3 · NEUTRAL INDICATOR
4 · INDICATOR WARNING LAMP
5 · HIGH BEAM INDICATOR
6 · OIL PRESSURE LAMP
7 · TACHOMETER LAMP

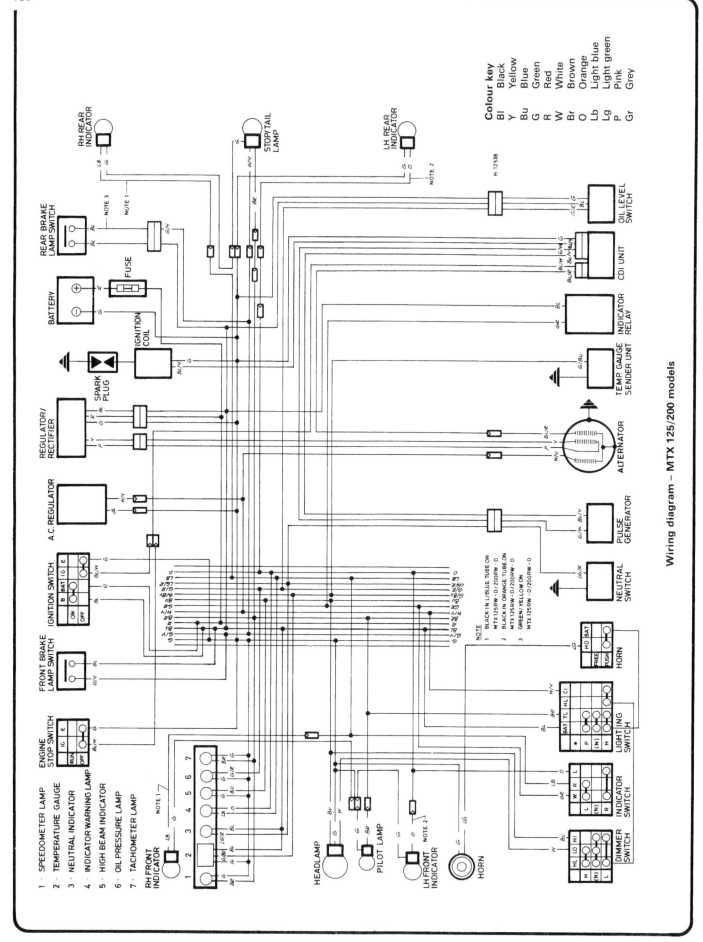

Wiring diagram – MTX 125/200 models

Conversion factors

Length (distance)
Inches (in)	X	25.4	= Millimetres (mm)	X 0.0394	= Inches (in)
Feet (ft)	X	0.305	= Metres (m)	X 3.281	= Feet (ft)
Miles	X	1.609	= Kilometres (km)	X 0.621	= Miles

Volume (capacity)
Cubic inches (cu in; in³)	X	16.387	= Cubic centimetres (cc; cm³)	X 0.061	= Cubic inches (cu in; in³)
Imperial pints (Imp pt)	X	0.568	= Litres (l)	X 1.76	= Imperial pints (Imp pt)
Imperial quarts (Imp qt)	X	1.137	= Litres (l)	X 0.88	= Imperial quarts (Imp qt)
Imperial quarts (Imp qt)	X	1.201	= US quarts (US qt)	X 0.833	= Imperial quarts (Imp qt)
US quarts (US qt)	X	0.946	= Litres (l)	X 1.057	= US quarts (US qt)
Imperial gallons (Imp gal)	X	4.546	= Litres (l)	X 0.22	= Imperial gallons (Imp gal)
Imperial gallons (Imp gal)	X	1.201	= US gallons (US gal)	X 0.833	= Imperial gallons (Imp gal)
US gallons (US gal)	X	3.785	= Litres (l)	X 0.264	= US gallons (US gal)

Mass (weight)
Ounces (oz)	X	28.35	= Grams (g)	X 0.035	= Ounces (oz)
Pounds (lb)	X	0.454	= Kilograms (kg)	X 2.205	= Pounds (lb)

Force
Ounces-force (ozf; oz)	X	0.278	= Newtons (N)	X 3.6	= Ounces-force (ozf; oz)
Pounds-force (lbf; lb)	X	4.448	= Newtons (N)	X 0.225	= Pounds-force (lbf; lb)
Newtons (N)	X	0.1	= Kilograms-force (kgf; kg)	X 9.81	= Newtons (N)

Pressure
Pounds-force per square inch (psi; lbf/in²; lb/in²)	X	0.070	= Kilograms-force per square centimetre (kgf/cm²; kg/cm²)	X 14.223	= Pounds-force per square inch (psi; lbf/in²; lb/in²)
Pounds-force per square inch (psi; lbf/in²; lb/in²)	X	0.068	= Atmospheres (atm)	X 14.696	= Pounds-force per square inch (psi; lbf/in²; lb/in²)
Pounds-force per square inch (psi; lbf/in²; lb/in²)	X	0.069	= Bars	X 14.5	= Pounds-force per square inch (psi; lbf/in²; lb/in²)
Pounds-force per square inch (psi; lbf/in²; lb/in²)	X	6.895	= Kilopascals (kPa)	X 0.145	= Pounds-force per square inch (psi; lbf/in²; lb/in²)
Kilopascals (kPa)	X	0.01	= Kilograms-force per square centimetre (kgf/cm²; kg/cm²)	X 98.1	= Kilopascals (kPa)

Torque (moment of force)
Pounds-force inches (lbf in; lb in)	X	1.152	= Kilograms-force centimetre (kgf cm; kg cm)	X 0.868	= Pounds-force inches (lbf in; lb in)
Pounds-force inches (lbf in; lb in)	X	0.113	= Newton metres (Nm)	X 8.85	= Pounds-force inches (lbf in; lb in)
Pounds-force inches (lbf in; lb in)	X	0.083	= Pounds-force feet (lbf ft; lb ft)	X 12	= Pounds-force inches (lbf in; lb in)
Pounds-force feet (lbf ft; lb ft)	X	0.138	= Kilograms-force metres (kgf m; kg m)	X 7.233	= Pounds-force feet (lbf ft; lb ft)
Pounds-force feet (lbf ft; lb ft)	X	1.356	= Newton metres (Nm)	X 0.738	= Pounds-force feet (lbf ft; lb ft)
Newton metres (Nm)	X	0.102	= Kilograms-force metres (kgf m; kg m)	X 9.804	= Newton metres (Nm)

Power
Horsepower (hp)	X	745.7	= Watts (W)	X 0.0013	= Horsepower (hp)

Velocity (speed)
Miles per hour (miles/hr; mph)	X	1.609	= Kilometres per hour (km/hr; kph)	X 0.621	= Miles per hour (miles/hr; mph)

Fuel consumption*
Miles per gallon, Imperial (mpg)	X	0.354	= Kilometres per litre (km/l)	X 2.825	= Miles per gallon, Imperial (mpg)
Miles per gallon, US (mpg)	X	0.425	= Kilometres per litre (km/l)	X 2.352	= Miles per gallon, US (mpg)

Temperature

Degrees Fahrenheit = ($^\circ$C x 1.8) + 32

Degrees Celsius (Degrees Centigrade; $^\circ$C) = ($^\circ$F - 32) x 0.56

*It is common practice to convert from miles per gallon (mpg) to litres/100 kilometres (l/100km), where mpg (Imperial) x l/100 km = 282 and mpg (US) x l/100 km = 235

Metric conversion tables

Inches	Decimals	Millimetres
1/64	0.015625	0.3969
1/32	0.03125	0.7937
3/64	0.046875	1.1906
1/16	0.0625	1.5875
5/64	0.078125	1.9844
3/32	0.09375	2.3812
7/64	0.109375	2.7781
1/8	0.125	3.1750
9/64	0.140625	3.5719
5/32	0.15625	3.9687
11/64	0.171875	4.3656
3/16	0.1875	4.7625
13/64	0.203125	5.1594
7/32	0.21875	5.5562
15/64	0.234375	5.9531
1/4	0.25	6.3500
17/64	0.265625	6.7469
9/32	0.28125	7.1437
19/64	0.296875	7.5406
5/16	0.3125	7.9375
21/64	0.328125	8.3344
11/32	0.34375	8.7312
23/64	0.359375	9.1281
3/8	0.375	9.5250
25/64	0.390625	9.9219
13/32	0.40625	10.3187
27/64	0.421875	10.7156
7/16	0.4375	11.1125
29/64	0.453125	11.5094
15/32	0.46875	11.9062
31/64	0.484375	12.3031
1/2	0.5	12.7000
33/64	0.515625	13.0969
17/32	0.53125	13.4937
35/64	0.546875	13.8906
9/16	0.5625	14.2875
37/64	0.578125	14.6844
19/32	0.59375	15.0812
39/64	0.609375	15.4781
5/8	0.625	15.8750
41/64	0.640625	16.2719
21/32	0.65625	16.6687
43/64	0.671875	17.0656
11/16	0.6875	17.4625
45/64	0.703125	17.8594
23/32	0.71875	18.2562
47/64	0.734375	18.6531
3/4	0.75	19.0500
49/64	0.765625	19.4469
25/32	0.78125	19.8437
51/64	0.796875	20.2406
13/16	0.8125	20.6375
53/64	0.828125	21.0344
27/32	0.84375	21.4312
55/64	0.859375	21.8281
7/8	0.875	22.2250
57/64	0.890625	22.6219
29/32	0.90625	23.0187
59/64	0.921875	23.4156
15/16	0.9375	23.8125
61/64	0.953125	24.2094
31/32	0.96875	24.6062
63/64	0.984375	25.0031

Millimetres to Inches	
mm	Inches
0.01	0.00039
0.02	0.00079
0.03	0.00118
0.04	0.00157
0.05	0.00197
0.06	0.00236
0.07	0.00276
0.08	0.00315
0.09	0.00354
0.1	0.00394
0.2	0.00787
0.3	0.01181
0.4	0.01575
0.5	0.01969
0.6	0.02362
0.7	0.02756
0.8	0.03150
0.9	0.03543
1	0.03937
2	0.07874
3	0.11811
4	0.15748
5	0.19685
6	0.23622
7	0.27559
8	0.31496
9	0.35433
10	0.39370
11	0.43307
12	0.47244
13	0.51181
14	0.55118
15	0.59055
16	0.62992
17	0.66929
18	0.70866
19	0.74803
20	0.78740
21	0.82677
22	0.86614
23	0.09551
24	0.94488
25	0.98425
26	1.02362
27	1.06299
28	1.10236
29	1.14173
30	1.18110
31	1.22047
32	1.25984
33	1.29921
34	1.33858
35	1.37795
36	1.41732
37	1.4567
38	1.4961
39	1.5354
40	1.5748
41	1.6142
42	1.6535
43	1.6929
44	1.7323
45	1.7717

Inches to Millimetres	
Inches	mm
0.001	0.0254
0.002	0.0508
0.003	0.0762
0.004	0.1016
0.005	0.1270
0.006	0.1524
0.007	0.1778
0.008	0.2032
0.009	0.2286
0.01	0.254
0.02	0.508
0.03	0.762
0.04	1.016
0.05	1.270
0.06	1.524
0.07	1.778
0.08	2.032
0.09	2.286
0.1	2.54
0.2	5.08
0.3	7.62
0.4	10.16
0.5	12.70
0.6	15.24
0.7	17.78
0.8	20.32
0.9	22.86
1	25.4
2	50.8
3	76.2
4	101.6
5	127.0
6	152.4
7	177.8
8	203.2
9	228.6
10	254.0
11	279.4
12	304.8
13	330.2
14	355.6
15	381.0
16	406.4
17	431.8
18	457.2
19	482.6
20	508.0
21	533.4
22	558.8
23	584.2
24	609.6
25	635.0
26	660.4
27	685.8
28	711.2
29	736.6
30	762.0
31	787.4
32	812.8
33	838.2
34	863.6
35	889.0
36	914.4

Index